T0211670

THE WORK OF THE SUN

THE WORK OF THE SUN

LITERATURE, SCIENCE, AND POLITICAL ECONOMY, 1760–1860

Ted Underwood

First published in 2005 by
PALGRAVE MACMILLAN™
175 Fifth Avenue, New York, N.Y. 10010 and
Houndmills, Basingstoke, Hampshire, England RG21 6XS
Companies and representatives throughout the world.

PALGRAVE MACMILLAN is the global academic imprint of the Palgrave Macmillan division of St. Martin's Press, LLC and of Palgrave Macmillan Ltd. Macmillan® is a registered trademark in the United States, United Kingdom and other countries. Palgrave is a registered trademark in the European Union and other countries.

ISBN 978-1-349-52928-5   ISBN 978-1-4039-8190-5 (eBook)
DOI 10.1007/978-1-4039-8190-5
Library of Congress Cataloging-in-Publication Data

Underwood, T. L. (Ted L.)
     The work of the sun : literature, science, and economy, 1760–1860 / Ted Underwood.
          p. cm.
     Includes bibliographical references.

     1. English literature—19th century—History and criticism. 2. Sun—In literature. 3. Politics and literature—Great Britain—History—19th century. 4. Politics and literature—Great Britain—History—18th century. 5. Literature and science—Great Britain—History—19th century. 6. Literature and science—Great Britain—History—18th century. 7. English literature—18th century—History and criticism. I. Title.

PR468.383U53 2005
820.9'36—dc22   2005048826

A catalogue record for this book is available from the British Library.

Design by Newgen Imaging Systems (P) Ltd., Chennai, India.

First edition: October 2005

10 9 8 7 6 5 4 3 2 1

Transferred to Digital Printing 2011

CONTENTS

ACKNOWLEDGMENTS

This book has taken a long time to write; in the process I have accumulated many debts. Conversations with Reeve Parker, Peter Dear, Mark Seltzer, and Harry Shaw inspired the project. Important leads were contributed by M. H. Abrams, Rob Anderson, Aimée Boutin, Jennifer Dakin, Jim Hansen, Nate Johnson, Karen Karbiener, Jeff Kasser, Anne Mallory, Laurie Osborne, Anindyo Roy, Elizabeth Sagaser, Katherine Stubbs, and David Suchoff. Readers, reviewers, and respondents who made a difference include Alan Bewell, Marshall Brown, Morris Eaves, Catherine Gallagher, Peter Manning, Robert Markley, Eric G. Wilson, and Gillen Wood. My editor, Marilyn Gaull, provided patient advice and encouragement. Farideh Koohi-Kamali and Lynn Vande Stouwe supervised production; Maran Elancheran corrected the manuscript.

Work on this book was supported by the Andrew W. Mellon Foundation through the English Department of Cornell University. Colby College supported research travel to the Widener and Huntington libraries. The Mario Einaudi Center for International Studies supported travel to London, where Mrs. I. McCabe of the Royal Institution assisted me with Humphry Davy's notebooks. The final stages of manuscript preparation were supported by the Research Board of the University of Illinois at Urbana-Champaign. The editors of *Modern Language Quarterly* have kindly permitted me to reprint a section of chapter 6 that originally appeared as "The Science in Shelley's Theory of Poetry," *MLQ* 58 (1997).

Eleanor Courtemanche has been my constant guide to the nineteenth century. This book is dedicated to her.

ABBREVIATIONS

Italics are in the source unless indicated as mine. Translations of French sources are mine unless otherwise attributed.

BP   William Blake, *The Complete Poetry and Prose*, ed. David V. Erdman (Berkeley: University of California Press, 1982).

BW   George Gordon Byron, *The Complete Poetical Works*, ed. Jerome McGann, 6 vols. (Oxford: Oxford University Press, 1986).

DW   *The Collected Works of Humphry Davy*, ed. John Davy, 9 vols. (London: Smith, Elder & Co., 1839).

Ex.   William Wordsworth, *The Excursion, Poetical Works*, ed. E. de Selincourt and Helen Darbishire, vol. 5 (Oxford: Oxford University Press, 1966).

GW   William Godwin, *Political and Philosophical Writings*, 7 vols. (London: Pickering, 1993).

LY   *The Letters of William and Dorothy Wordsworth: The Later Years*, ed. Alan G. Hill and Ernest de Selincourt, 2nd ed., 2 vols. (Oxford: Oxford University Press, 1978).

MY   *The Letters of William and Dorothy Wordsworth: The Middle Years*, ed. Ernest de Selincourt, Mary Moorman, and Alan G. Hill, 2 vols. (Oxford: Oxford University Press, 1970).

1805   William Wordsworth, *The Prelude: 1799, 1805, 1850*, ed. Jonathan Wordsworth, M. H. Abrams, and Stephen Gill (New York: W. W. Norton, 1979). All references to *The Prelude* are to the 1805 text.

PrW   William Wordsworth, *Prose Works*, ed. W. J. B Owen and Jane Worthington Smyser, 3 vols. (Oxford: Oxford University Press, 1974).

PW   William Wordsworth, *Poetical Works*, ed. E. de Selincourt and Helen Darbishire, 2nd ed., 5 vols. (Oxford: Oxford University Press, 1965–1966).

RP   Mary Robinson, *The Poetical Works*, 3 vols. (London: Richard Phillips, 1806).

SP   Percy Bysshe Shelley, *Poetical Works*, ed. Thomas Hutchinson, new edition corrected by G. M. Matthews (London: Oxford University Press, 1970).

SPr   Percy Bysshe Shelley, *Shelley's Prose; or, The Trumpet of a Prophecy*, ed. David Lee Clark (London: Fourth Estate, 1988).

Task William Cowper, *The Task, The Poems of William Cowper*, ed. J. D. Baird and C. Ryskamp, vol. 2 (Oxford: Oxford University Press, 1995).

ILLUSTRATIONS

INTRODUCTION

"The source of all labour is the sun," *Chambers's Journal* informed its readers in 1866. "All the labour done under the sun is really done by it." The glide from "done under" to "done by" concisely dramatizes the appeal of this claim, which was by the 1860s a commonplace of British journalism. The light revealing the world and the labor shaping it are two aspects of a single protean power. The sun, as the source of that power, binds the natural order to the order of economic production—at once personifying natural force as a worker, and elevating work to an ontological principle. In the words of the scientific lecturer John Tyndall: "every mechanical action on the earth's surface, every manifestation of power, organic and inorganic, vital and physical, is produced by the sun. . . . He builds the forest and hews it down, the power which raised the tree, and which wields the axe, being one and the same."[1]

These paeans on the sun exemplify Victorian writers' broader fascination with theories that correlate work and natural force, and thereby idealize work as participation in the life of nature. Anson Rabinbach has named this representation of work *productivism*: "the belief that human society and nature are linked by the primacy and identity of all productive activity, whether of laborers, of machines, or of natural forces." A productivist interpretation of work undergirds much nineteenth- and early-twentieth-century social theory. Analogies between natural force and human labor power figured importantly in Marxism, but also in new sciences of efficiency celebrated by Right and Left alike; they fostered anxieties about agency and identity in late-nineteenth-century fiction.[2]

The emergence of the productivist analogy between natural force and work remains a historical puzzle. Rabinbach calls it a consequence of "the conceptual revolution ushered in by nineteenth-century scientific discoveries, especially thermodynamics."[3] But though analogies between work and natural force eventually took a thermodynamic form, thermodynamic science did not give rise to productivism. The first law of thermodynamics (the principle of energy conservation) developed out of experiments conducted in the 1840s; it was expressed formally by Hermann von Helmholtz in 1847. Theories of production that linked society to nature by correlating work and natural force developed earlier, in the first three decades of the nineteenth century. They already played an important role in debates over trade protection—for instance in the dispute over the Corn Laws in Regency Britain, where the notion that industry cooperated with the forces of nature was used to argue that manufacturing enterprises were as natural, as productive, and as deserving of protection as agriculture. By 1829, both Charles Babbage and G. G. Coriolis, working independently, had quantified the relationship between the "work" of nature and of the machine or human laborer. Their arguments were not thermodynamic, nor would it be accurate even to say that they depended on "nineteenth-century scientific discoveries." The measurements that Babbage and Coriolis used had been developed by eighteenth-century engineers; what the nineteenth century added was a clear expression of their economic significance.

The analogy between human labor and natural force was not, on the other hand, intuitive or time-honored. As recently as the seventeenth century, work had seemed a distinctively human assertion of moral will. Far from an expression of nature's power, it was a struggle against the idleness and inertia of the non-human world, which only became productive when worked and turned by the plow. Even in Eden before the Fall, according to John Milton,

> Man hath his daily work of body and mind
> Appointed, which declares his Dignitie
> And the regard of Heav'n on all his waies;
> While other animals unactive range
> And of thir doings God takes no account.[4]

In this Protestant context, work was not a curse but a mark of dignity that distinguished the human realm from the animal and vegetable creations. Only humans had the capacity to work, because work was defined as the conscious and deliberate fulfillment of an "appointed" task.

By 1775, this conception of work was being revised. It seemed increasingly important for human work to resemble, or directly participate in, the spontaneous "energies" of nature. The analogy between work and natural energy had important political and cultural consequences several decades before it was expressed in mathematical terms. William Godwin compared the aristocratic suppression of middle-class talent to the "indiscriminate destruction" of nature's productive "energy."[5] Poets specialized in exploring the reciprocity between the creative powers of the human mind and of nature—focusing especially on the idea that the "self-sufficing power of solitude" in their own lives mirrored "the enduring majesty and power / Of independent nature" (*1805* 2:78, 8:785–86). Chemists and physicians sought to trace human energy to its natural source—finding that source notably in sunlight, transmuted by plants into the "dephlogisticated air" or "phosoxygen" that fueled consciousness and muscular motion (*DW* 2:86). *The Work of the Sun* explores the growing importance of analogies between work and natural force between 1760 and 1860. These analogies were certainly related to industrialization, which made motive power visible as a fungible commodity with a price of its own. But the connection to technologies of power was not at first direct; it was mediated both by class struggle and by the prestige of particular kinds of scientific explanation. Abstract concepts of "force" dominate eighteenth-century and Romantic writing, for example, partly because Isaac Newton insisted that explanations should be framed "not as occult Qualities, supposed to result from the specifick Forms of things, but as general Laws of Nature, by which the Things themselves are form'd."[6] Well into the nineteenth century, natural philosophers continued to debate whether the powers of nature were subtle elastic fluids or immaterial forces of attraction and repulsion. But the advocates of both approaches took it for granted that they were talking about agencies that operated equably everywhere, not attributes of specific entities or forms. That habit of separating agency from agents provided an important precedent for the naturalization of work.

The process of naturalization was more immediately driven, however, by the political restiveness of the middle classes in the eighteenth century. Traditional concepts of rank, order, and estate were giving way, as Stephen Wallech has observed, to a new concept of "class" that sorted individuals according to their mode of producing or consuming economic value. American and English conflicts over parliamentary representation had accordingly evolved, by the 1780s and 1790s, into what Isaac Kramnick calls "a self-conscious glorification of the mission of the middle class in English society." Philosophers like William Godwin and Mary Wollstonecraft argued that activity and independence (displayed most prominently in the "middle ranks of life") were vital principles supporting all social existence. Poets and novelists framed analogies between work and natural force in order to show that middle-class workers were autonomous and politically responsible citizens—with no need of the inherited land that republican tradition had taken as a prerequisite of material independence and virtuous citizenship.[7] Work itself was to provide an independence like the mobile and protean autonomy of light, heat, and electricity.

At this point, the class conflict motivating an idealization of middle-class work is still more closely related to the financial revolution of the late seventeenth and early eighteenth centuries than to steam engines or spinning jennies. The analogy between natural power and work did not yet encompass machines—indeed, an inclusion of machines or of unskilled "mechanic" labor would have negated the point of the analogy, which was to emphasize the economic independence of middle-class life. Nor did philosophers share any consensus about the way natural power should be measured or compared to human labor; one should not read implicitly quantitative conceptions of energy and force into contexts where no dimension of measurement is envisioned. But the analogy between human and natural power was already something more than a normative metaphor: many poets and physicians claimed that human energy and ambition not only resembled natural force in their spontaneity, but were produced physiologically by nature's action on the mind—in particular by sunlight.

In the early nineteenth century, class conflicts were articulated as an economic debate between landed and manufacturing interests, and linked to reflection on mechanical production. By the late 1820s, engineers and political economists had quantified the analogy between work and natural force, and had extended it to machines and unskilled labor—in the process weakening its connection to the claims of autonomy made by middle-class artisans, professionals, and intellectuals. The analogy became a way of idealizing industrial production in general—a politically significant gesture, to be sure, but one that no longer consistently served the political interests of a single class. The railway engineer George Stephenson held a productivist view of work, but so did Friedrich Engels.

This book considers texts of several different kinds—including works of political economy, engineering theory, medicine, fiction, and popular journalism. But I am particularly interested in showing how the developing analogy between work and natural force affected British poetry between 1780 and 1830. (Following recent practice in British literary studies, I use the adjective "Romantic"—with a capital R—to designate this period of literary history, rather than a stylistic emphasis that cuts across several different periods.) Many critical observations about Romantic poetry can be traced to writers' habit of interpreting their experience as a manifestation of unattributed power. A vocabulary of abstract agency is prominent first of all in poets' description of their literary labor. Poetry is less a skill residing in the poet, or a quality of the formed poem, than a power uniting the poet and poem to the activity of the universe. William Hazlitt's lecture "On Poetry in General" is typical. For Hazlitt, poetry exists "wherever there is a sense of beauty, of power, or harmony, as in the motion of a wave of the sea [or] in the growth of a flower that 'spreads its sweet leaves to the air, and dedicates its beauty to the sun.'"[8] Similar descriptions of poetry as a fungible power latent in all perception (or indeed in all life) appear in Mary Robinson, William Wordsworth, S. T. Coleridge, and P. B. Shelley.

Romantic writers' fascination with abstract agency is equally prominent in their descriptions of landscape. If the speakers of Romantic poems sometimes forget themselves in precise observation of a physical scene, they also conversely reduce nature to a screen or veil for placeless power, whether that power is understood as the speaker's own transforming imagination, or as an impersonal "secret Strength of things" (*SP* 535). Critics have variously interpreted this hesitation between particularity and abstraction as an effect of philosophical skepticism (in P. B. Shelley), or of a poetics of sensibility (in Mary Robinson), or (in William Wordsworth) as the stance of a mind moved by natural objects to a meditative self-consciousness that reaches beyond them.[9] These critical descriptions are sensitive and accurate, but the homology between them suggests that writers' responses were also shaped by broader and less conscious structures of feeling. In this book I argue that Romantic writers' recourse to the vocabulary of abstract agency was, as much as anything else, a fantasy about economic autonomy. For a few years in the 1790s, that fantasy became overtly utopian and politically charged, because *energy* was a code word for middle-class radicals' faith in the talent and activity of their own class, and a target of satire for conservatives who doubted that sheer activity was any proof of political responsibility. But the habit of invoking dynamic abstractions had a latent social significance both before and after the 1790s, although the gesture was not narrowly limited to a particular party. When writers pointed to energy as a sign of personal independence, they tacitly neglected an older rhetorical gesture that would have pointed to a well-ordered estate. When poets aspired to possess "a power like one of Nature's" (the phrase is Wordsworth's, but other writers thought about literary ambition in similar terms), they defined literature as a human reflection of the spontaneity revealed in natural force (*1805* 12:312). In poetry, as elsewhere, the analogy tends to implicitly endorse a middle-class conception of independence—mobile, protean, unencumbered by land.

Wordsworth is admittedly better-known for "agrarian idealism"; some of his poems dwell lovingly on particular plots of ground, and purport to show that landed property is the necessary stay of domestic morality and civic independence. I acknowledge a real tension here.[10] Some Romantic-era writers (William Cowper, for instance, as well as Wordsworth) certainly did distrust commercial society, and embraced georgic imagery associated with an older model of rural autonomy. But even those writers were attracted by a rhetoric of dynamic abstraction that dissolves the agrarian vision into something more portable. Cowper's speaker in *The Task* looks through the apparent stability of the landscape to a restless undulation that sustains nature, and concludes that security can be found only in "ceaseless action." Wordsworth's Michael is slowly pried away from his paternal hills and fields, and finds that he can finally rely on nothing but wind and sun, and the work of his own hands (*Task* 1:367, *PW* 2:93–94).

Romantic analogies between work and natural force do not usually celebrate mechanical production directly. But by uniting different kinds of work (agricultural, poetic, artisanal) in a single commutable power, and by representing that productive power (rather than land) as the link between nature and culture, Romantic poets diffuse the special significance agriculture had once possessed into a larger productive system. For Adam Smith, the cultivated field was still a privileged site of cooperation between the productive powers of nature and of man.[11] But early-nineteenth-century poets (like early-nineteenth-century political economists) insist on the ubiquity of that cooperation. Victorian lecturers like John Tyndall, and industrialists like George Stephenson, carry the process further, reimagining agriculture as part of a explicitly industrial metabolism of power. Romantic poets and Victorian industrialists identified their endeavors with natural force in different ways, but for similar reasons; both groups were engaged in redistributing the symbols of agriculture's foundational economic role to other occupations and classes.

A reading of Romantic poets that stresses their affinity with industrialists may seem counterintuitive, but it need not deny the genuine critical impulses of Romantic poetry. As Raymond Williams stresses, Romantic writers use the interplay of light and air and leaves to imagine a cooperative community missing from their own economic experience. The "green language" of nature offers a refuge from "what man has made of man" (*PW* 4:58). But Williams rightly goes on to concede that these sympathetic impulses are dramatized with "a new emphasis . . . on the dispossessed, the lonely wanderer, the vagrant. . . . Thus an essential isolation and silence and loneliness have become the only carriers of nature and community against the rigours, the cold abstinence, the selfish ease of ordinary society."[12] In this book I stress that the paradox Williams locates in the dramatic situation of Romantic poetry is also built into the green language itself. The habits of thought that allow Romantic writers to see twigs and bare hills as an interplay of vital forces also allow those writers to compress the entire visible scene into an emblem of productive power. Romantic writers use nature both to flee from, and to represent, the rigors of economic individualism.

One way to grasp the significance of this thesis is to put it face to face with Leo Marx's *Machine in the Garden*, a classic and still useful study of the relationship between nineteenth-century technology and literature. Marx was one of the first scholars to notice that early-nineteenth-century writers often fail to display the expected tension between pastoral ideals and industrial progress. Many American writers—from Tench Coxe to Ralph Waldo Emerson—confidently welcomed the steam engine as a power that would transform the wilderness into a garden of prosperity and independence. I should remark in passing that this is not a uniquely American phenomenon—as chapter 5 of this book will indicate. But I am at present more interested in the second half of Marx's argument. There he traces a growing American disillusionment with technology, expressed (on the canvasses of Thomas Cole and in the narratives of Nathaniel Hawthorne and Herman Melville) as "a sense of the opposition between industrial fire and pastoral sun." Fire is in this scheme an emblem of Satanic hubris; painters and writers pose sunlight against it to evoke a lost pastoral Eden.[13] Marx's readings are entirely persuasive, but an interesting irony emerges when those readings are juxtaposed with the thesis of the present study—which is that nineteenth-century sun-worship often translates agrarian visions of social order into a commercial or industrial form. Romantic idealizations of natural force are shaped by the same fascination with placeless, impersonal autonomy that led John Tyndall to declare, "there is not a hammer raised, a wheel turned, or a shuttle thrown, that is not raised, and turned, and thrown by the sun."[14] Writers who criticized industrial hubris by contrasting it to a "pastoral sun" were thus actually exploring the tension between two different aspects of an industrial ideal; their pastoral sunlight is largely a reflection of industrial fire.

Readers with ecocritical interests may find this a depressingly cynical thesis. I want to stress that I do not mean it cynically. I like that unpopular Romantic abstraction, "nature," and I think a fuller awareness of its economic dimension will actually deepen readers' enjoyment of Romantic writing. Many Romantic writers, after all, emphasized that nature borrows solemnity from the human eye, and even—in the words of "Michael"—from the observer's expectation "of honourable gain" (PW 2:82). Romantic affection for nature will survive being interpreted as longing for material security. Admittedly, I argue that Romantic nature poetry often envisions a sort of security apposite to commercial and industrial rather than agrarian life. Since some writers' overt politics leaned the other direction, this implies an ambivalence. But the ambivalence need not be understood as a weakness; it makes the poetry livelier and more candid. Early-nineteenth-century poetry and political economy do often make exaggerated promises of independence. David Ricardo wants to believe that manufacturers tap a natural power that is "inexhaustible, and at every man's disposal"; Wordsworth wants to believe that pedlars and poets share the invulnerable ubiquity of roadside flowers.[15] There is a strong element of fantasy in both discourses: a fantasy of absolute independence that would be cruel if taken literally, and that needs to be read with the same awareness of distortion that Freud asks readers to apply to dreams. But to understand nature writing as a dream about material security is not therefore to despise it.

The following pages give a connected chapter-by-chapter synopsis of the book's argument—intended, frankly, as a guide for skimmers. Since this book touches on many different disciplines, I cannot expect all readers to be equally interested in all sections of the argument. In particular, readers who are interested in literature (but not in eighteenth-century science) should probably start with the second chapter.

Chapter 1 describes the emergence of sunlight as a figure for fungible agency in eighteenth-century natural philosophy. There was nothing new about the assumption that the sun somehow fostered life. Early modern writers had described sunlight as a taxonomic power that differentiated terrestrial matter to produce the characteristic forms and virtues of living things. In the latter half of the eighteenth century, natural philosophers diverged from that tradition by defining sunlight, not as a power that produced specific forms, but as a power of vital agency abstracted from form—a power that caused all matter to move, to perceive, to respond to stimuli. Sunlight thus became the leading example of an agency that could be measured without reference to the forms or powers of discrete agents. French chemists even sought to discover the ratio of conversion between sunlight, oxygen, and human work.[16]

Chapter 2 moves from natural philosophy to politics, examining the resonance of the word *energy* in the political philosophy and fiction of the 1790s. For many writers in this period, the ideal of free and spontaneous human energy was a way of harnessing natural philosophy to middle-class economic and political interests. Praise of "great energies" became a tic in William Godwin's prose, for instance, because the phrase concisely bound together several of his central beliefs—especially his claim that great talents are like natural forces, welling up spontaneously only to be wasted by a corrupt aristocratic society. *Energy* became such a buzzword in radical prose that conservative opponents fastened on the word itself as a target for parody, linking it to a caricature of radical philosophers as libertine arrivistes. In this chapter I also explain my reasons for using the hoary term "class" to analyze social change between 1780 and 1840. I realize that there was no homogenous urban bourgeoisie in late-eighteenth-century England. I also realize that the middle class has been rising for hundreds of years, and is exhausted. If I could honestly attribute the changes discussed here to "professionalization," I would happily use that fresher word instead. Unfortunately, different professions felt little common cause in the 1790s, whereas it is easy to document self-conscious advocacy of "the middle ranks of life." Class is, for this moment in history, exactly the right analytic category.

Chapter 3 shows how the themes considered in the two previous chapters began to fuse in British Romantic poetry. Since late-eighteenth-century philosophers used sunlight to represent nature as a mobile and formless force, and the spontaneity of natural force was closely associated with middle-class ambition, it was logical to conclude that liberty and enterprise animate society just as the sun animates the terrestrial world. Mary Robinson implies as much in her poem on "The Progress of Liberty." Romantic poets' well-known fascination with solar gods—from Prometheus to Apollo to the Persian deity Mithras—was founded on a similar chain of inferences. Because the sun was understood to preside over liberty in general, and middle-class aspiration in particular, poets increasingly invoked Apollo rather than a Muse as the patron of poetic ambition. Certain poets who were also natural philosophers (for instance, Humphry Davy) took this connection quite literally, representing the sun's rays as physical fuel for aspiring minds. Writers less directly immersed in natural philosophy (for instance, the young John Keats) took the connection less literally, but still used solar mythography to represent ambition as a natural force. Chapter 3 closes by distinguishing William Blake's sun-god Los from the main current of solar mythography in the 1790s. In spite of his affinity for "energy," Blake generally avoids the analogy between work and natural force traced in this book, instead holding fast to the premise of indivisible personal agency. So though Los is a sun-god and a god of work, he never becomes (like the Romantic Apollo) an emblem of the continuity between human ambition and the powers of nature. Blake diverged here from other writers because he associated the divisibility of natural force with the division of labor—which posed a more serious economic problem for artisans than for professional and clerical workers.

Since scientists and poets often used the sun as a synecdoche for natural force in general, several chapters of this study use that fact to organize the discussion. But this book is not a study of solar imagery. Romantic writers sometimes used electricity in a similar way, or invoked plural and vaguely-defined energies to acknowledge their own ignorance. What matters for this book's argument is not a particular force, but the underlying assumption that formless, placeless forces are responsible for the multiform phenomena of physical and social existence. The ubiquity of sunlight made it a handy emblem for that model of agency, but writers' insistence on condensing unlocatable agency into a solar orb or sun-god also reflected some discomfort with the (relatively new) idea being represented. As G. W. F. Hegel put it, the sun is an individuality privileged to signify universality—a figure that rapidly becomes self-contradictory if taken literally.[17] Some Romantic writers (notably Wordsworth) were troubled enough by this contradiction that they carefully avoided condensing nature's power into any positive presence. In short, while many Romantic writers used the sun to focus the analogy between natural force and work, that analogy was also inherently resistant to figuration, and the writers who thought hardest about it some-times eschewed solar imagery.

Chapters 1 through 3 have outlined the emergence of a new way of using the nonhuman world to represent economic independence. But this change did not go unchallenged. Several poets who were fascinated by the analogy between natural force and independence also resisted political identification with the middle ranks of life. Chapters 4 and 5 therefore pause to consider conflicts between the emerging middle-class conception of independence as spontaneous work and an older conception of auton-omy founded on possession of land. I focus particularly on William Cowper and William Wordsworth. For reasons of elective affinity as much as actual class origin, both writers were positioned uneasily between the professional middle classes and the landed gentry. Both were forced to compromise between middle-class and aristocratic ways of connecting culture to nature.

Chapter 4 focuses on Cowper, and discusses the fate of the georgic poem at the end of the eighteenth century. Some literary mode had always been needed to justify human labor in the context of a larger natural order. For classical antiquity, and for much of the eighteenth century, the georgic mode filled this role. But where classical georgics had represented work as a contest between human will and natural indolence, eighteenth-century georgics increasingly represent work as a mode of participation in a natural order that is itself a ceaseless worker. In William Cowper's long poem *The Task* (1785), this tendency reaches its logical extreme, and the georgic poem as such ceases to exist. Human work is assimilated to the restless undulation of nature, with the result that the work held up for praise is no longer delib-erate and industrious, but spontaneous. The ideal of spontaneous work allowed Cowper to resolve a social dilemma that was peculiarly pressing in his own case: a conflict between Protestant celebration of work and genteel retirement. Cowper does not resolve the conflict neatly: his spontaneous work can look like idle rambling, and it remains shadowed by guilt. But *The Task* embraces that consciousness of errancy as a hopeful sign of moral and physical struggle.

Wordsworth's *Excursion* resembles *The Task* in several ways. Wordsworth too is uneasy about the georgic ideal of retirement (represented in Wordsworth's poem by a despairing Solitary).[18] And as the title of *The Excursion* might suggest, Wordsworth too celebrates errancy (imaginative and physical) as a way of embracing social contradictions he cannot per-fectly resolve. The similarities run deeper still, because the contradictions explored in *The Excursion* are also rooted in a conflict between the ideals of the landed gentry and of the middle classes. As a land-agent's son, renting Dove Cottage and hoping for a long-deferred inheritance, Wordsworth belonged to the professional middle class—but to a portion of that class

whose interests were closely bound up with those of the gentry. Quite apart
from his family connections, Wordsworth was intellectually committed to
an agrarian model of community that represented civic responsibility as
ownership and physical occupation of land. This too linked him to aristocratic ideals.[19] Wordsworth's poems nevertheless display a strong sense of
peripatetic vocation, revealing the author's (slightly regretful) consciousness
that his work as a poet must finally be classed with the work of lawyers, pedlars, and other entrepreneurs who carry their capital in a bag (or in their head).

In chapter 5, I argue that this sense of itinerant vocation waxes strong
enough in *The Excursion* to undermine Wordsworth's overtly agrarian politics, and to foster a surprisingly intense sympathy with industrial power.
A cotton mill is of course a stationary thing, but *The Excursion* interprets
industry as a restless principle, whose mobility is revealed in the ceaseless
respirations of commerce and in the tides of workers who stream in and out
of the mills each night, "While all things else are gathering to their homes"
(*Ex.* 8:159). This restlessness is in part a nightmare: Wordsworth is fully
aware of the confinement and monotony of factory labor. But *The Excursion*
also locates a characteristically Wordsworthian security in the midst of disruption. It does so by interpreting industrial production as a natural force,
and stressing the ubiquity of that force. Industrial engines, like pedlars or
poets, are not bound to a particular estate; they can make themselves at
home wherever there are hills and streams, air and water. Wordsworth's
pedlar hero, the Wanderer, explicitly compares this industrial mobility to
the mobility of an imaginative mind. So although *The Excursion* strongly
criticizes the factory system, it celebrates industrial engines themselves as
natural (if errant) powers. In doing so, it prefigures Ricardian political
economy: a few years later, in *Principles of Political Economy and Taxation*
(1817), David Ricardo would likewise celebrate industry by stressing its
connection to mobile natural force.

I have argued that the analogy between human enterprise and natural
force had in the late eighteenth century a middle-class provenance. That
analogy retained the impress of middle-class interests even when poets like
Cowper and Wordsworth sought to graft it onto an agrarian ideal. But in
the second and third decades of the nineteenth century, a younger
generation of writers expanded the analogy in new ways that did fundamentally revise its social significance. Identification with ubiquitous force
became increasingly central to Romantic poetry: Keats's "Sleep and Poetry,"
Shelley's "Hymn of Apollo," Clare's "Dawning of Genius," Hemans's
"Properzia Rossi," and the soliloquies of Byron's Manfred all hinge on the
speaker's identification with a formless power symbolized by sunlight. But
it becomes less and less useful to read this identification with natural force
as an expression of determinate class interests. The obstacle to that reading

is first of all that the symbols of middle-class aspiration have been appropriated, in the Regency period, by writers who nearly span the range of imaginable class positions: Byron was a baron, and Clare a Northhamptonshire farm-hand. At the same time, the range of aspirations actually evoked was grow-ing narrower. When Mary Robinson and William Godwin compared human "energy" to natural force, they were talking about a kind of active merit that could be applied to any career—or at least, to any career not absolutely mechanical. When similar imagery appears in Keats or Shelley, it tends to define a specifically cultural aspiration. Middle-class idealizations of autonomous work have been reinterpreted as descriptions of the autonomy of culture in particular. Chapter 6 traces this process, paying special attention to writers' reasons for comparing poetry to sunlight.

While poets developed analogies between work and natural force in one direction, political economists and engineering theorists were pursuing a different train of thought. Until the 1820s, with one important exception, natural philosophers had made no effort to discover an exact ratio between the vital energy consumed by the body and the work actually accomplished by laborers. The obstacle to these experiments was not a lack of measure-ment techniques: the one exception to the rule (the "Premier mémoire sur la respiration" published by Antoine Lavoisier and Armand Seguin) suggests that measurement was not an intractable problem. The obstacle—at least in Britain—was rather a lack of interest in the connection between spontaneous energy and the toil of conscripted laborers. The social function of the word *energy* was to define a kind of work that was the opposite of "mechanic toil," because it was spontaneous rather than forced. Energy might be deployed in bodily or mental action: that distinction was not crucial. But to link energy to constrained, repetitive acts such as pumping a pedal or turning a crank would destroy the very aspect of the concept that made it interesting to contemporaries, by making it "mechanic" in every sense of that loaded word.

Chapter 7 recounts the belated application of analogies between sunlight and human work to the realm of mechanic toil. This required less the inven-tion of a new hypothesis than the systematic restriction of an old one. Between 1829 and 1834, philosophers began to interpret the sun's respon-sibility for terrestrial life and energy more narrowly as a claim about motive power. Changes in engineering theory and controversy over the British Corn Laws focused attention on the economic importance of motive power, and especially on a particular dimension of measurement that Coriolis called *travail*, or work. John Herschel was quick to combine this measure-ment with older chemical arguments about sunlight and coal, in order to suggest that all terrestrial motive power was derived from the sun. Thomas Carlyle's *Sartor Resartus* popularized the argument in a less mathematical form, representing the sun as the origin of "coal force," of "iron force," and of "the force of man."[20]

I conclude by observing that the enthusiastic British reception of the first law of thermodynamics in the 1860s hinged on prior acceptance of these productivist ideas. For lay readers the important thing about the first law was not the conservation principle itself: in fact, popular accounts of the first law did a very poor job of explaining which quantity was conserved. Victorian readers were more interested to see science confirm that all terrestrial production—natural, human, and mechanical—had the same solar origin. This is why so many journalists and lecturers mistakenly gave Herschel, Carlyle, or George Stephenson credit for anticipating the first law: in their view, what had been discovered was less a conservation principle than a proof of the long-suspected connection between the light revealing the world and the labor shaping it.

CHAPTER 1

LIGHT AS FLUID AGENCY

A study of correlations between work and natural force might well begin in the 1820s, when G. G. Coriolis defined work mathematically. But to begin at that point would evade an interesting historical problem. Late-eighteenth-century physicians and philosophers began to assume that human work was commensurable with a vital power diffused through nature before they had settled on an appropriate dimension of measurement for work or for vital power. This chapter investigates the motives that prompted that hasty assumption. I begin by asking how eighteenth-century natural philosophers came to view light, in particular, as a common denominator that united different kinds of agency.

The assumption that the sun informed terrestrial life was by no means new. Seventeenth-century iatrochemists had represented sunlight as a power that differentiated matter by endowing each kind of being with its form and characteristic virtues. Other writers made the sun a source of motion: a macrocosmic heart, it moved celestial and terrestrial bodies by magnetic influence, or by pumping light through a vast circulatory cycle.[1] The eighteenth century's contribution was to prune this assortment of models down to a statement of direct equivalence between sunlight and agency. Philosophers ceased to rely on theories that made the sun a spinning magnet or a pump. Light was itself a power initiating motion, quite apart from any circulatory impetus given to it by the sun. The connection between light and life was similarly streamlined. By the end of the century, it seemed too bold to claim that the sun endowed living beings with discrete organizing principles, and too cautious to say that the sun merely warmed the world so that life could flourish. The sun diffused a power of "sensation" and "spontaneous motion" that constituted life. Sunlight itself was agency— a generalized power to perceive, to respond to stimuli, and to initiate movement. Instead of directly shaping surrounding worlds or sustaining their motion, the sun animated its system simply by diffusing light.

Instead of acting on the world mechanically (by rotation, for instance), or by impressing form, the sun now governed the world dynamically—by infusing it with formless force. Dynamic models drew their eighteenth-century prestige in large part from Isaac Newton's theory of gravitation, and from his instruction to describe natural causes "not as occult Qualities, supposed to result from the specifick Forms of things, but as general Laws of Nature, by which the Things themselves are form'd."[2] Following what they believed to be Newton's example, natural philosophers attempted to explain all phenomena by reference to opposed forces of attraction and repulsion. The apparently solid and stable forms of things were epiphenomena produced by a ceaseless struggle between these two forces, neither of which had any inherent connection to form. By the end of the eighteenth century, light was widely identified with the force of repulsion, partly because the period's radical empiricism made it plausible to argue that causes of sensation were also causes of existence, and partly because chemists were just then exploring light's role in the production of oxygen or "vital air." Light was clearly connected to heat and electricity; philosophers disagreed about the nature of these forces, and disagreed about which of the three was more fundamental. But whether they wrote about electricity or about light, and whether they imagined these agents as immaterial forces or as subtle elastic fluids, late-eighteenth-century natural philosophers agreed implicitly on a definition of their basic research project, which was to trace the agency expressed in animate and inanimate motions back to a few primitive dynamic laws.

That project had a social resonance extending well beyond natural philosophy. Light, heat, and electricity, after all, are not the only forms of power that transcend the boundaries of discrete agents to circulate through the world in fungible and convertible quantities. Money fascinates for a similar reason. As Michel Foucault observed in *The Order of Things*, more than one eighteenth-century discourse sought to analyze observable forms into their hidden generative principles.[3] In looking for a formless force (or a pair of forces) that produce all the "specifick Forms" of agency, eighteenth-century natural philosophers closely resemble the political economists who were at the same time trying to trace wealth from its instantiation in gold and other commodities back to its source in production.

It now seems possible to go cautiously a step or two beyond Foucault's explanation of this parallel. (He was, strictly speaking, less concerned to explain than to describe it.) The new emphasis on production in eighteenth-century economic discourse, for instance, seems to be bound up with the emergence of the new social category of class, which categorizes citizens not by social rank, but by the role they play in producing or consuming value.[4] Foucault also seems to understate the influence of empirical skepticism—which set out explicitly (for instance in Newton's advice quoted above) to dissolve all natural forms into formless generative principles. But in making these observations I only reframe the question. Why did empirical skepticism flourish in philosophy at the same time as the category of class began to replace rank, order, and estate? To answer that question it would be necessary to write a social history of empiricism in the seventeenth century—which lies outside the purview of this book. It seems better to begin by admitting and embracing the necessary incompleteness of any historical explanation. This study begins in an era when two parallel changes were already underway. Theorists of wealth increasingly sought to analyze the value of all commodities into abstract productive power (whether that power was thought to reside in land, in labor, or in both). At the same time, natural philosophers sought to analyze the diversity of worldly forms into a few formless generative forces. Between 1780 and 1840 those discourses converged to create a general philosophy of production; this book will describe the process of convergence.

The Sun's Informing Power in Decline

The sun's contribution to life had often been conceived to have two different aspects: "informing light and moving flame," as James Thomson put it.[5] The sun's light, to this way of thinking, informed beings that were then animated and set in motion by the sun's warmth. Late-eighteenth-century philosophers' emphasis on light's moving power thus involved, not the invention of a wholly new model, but the restriction and reapplication of an old one. Sunlight lost its power to inform, and became instead the preeminent example of "moving flame." This chapter will examine three factors that contributed to the shift: the conflation of Newton's theory of light with his theory of the æther, the rise of radically skeptical empiricism, and the discovery that plants produce "vital air" when exposed to sunlight.[6]

Seventeenth-century iatrochemical tradition had described light as the source of a ferment that differentiated gross matter.[7] The sun "informed" the world in the sense that it endowed all animate and inanimate bodies with their distinguishing virtues. Its differentiating power opened up taxonomic space just as its light opened up the space of vision. Sometimes the connection was relatively indirect; the Dutch chemist Hermann Boerhaave, for instance, said only that the sun fostered a process of fermentation that perfected each body's "true genius."[8] Other writers believed that sunlight actually entered into, and constituted, the identity of each being. Isaac Newton's early alchemical writings identify sunlight with a "vegetable spirit" that produces generation and growth in plants, animals, and metals.[9] The spirit causes change by joining intimately with each body to create an actively fermenting "seed" that is the body's specific principle of action, differentiating and assimilating formless matter to produce form.[10]

These alchemical writings remained unpublished, but the queries Newton added to later editions of the *Opticks* could be used to reach similar conclusions. Stephen Hales, for instance, quoted Newton's published speculation that "bodies receive much of their Activity from the Particles of Light that enter their Composition" in order to support his own theory that light particles "ennoble" vegetable juices, helping to produce "the most racy generous tastes of fruits, and the grateful odours of flowers."[11] Here light's contribution is still to endow the plant with an active juice that produces its distinguishing character. But in emphasizing light's particulate nature, Hales and Newton diverge from alchemical tradition. Light is no longer strictly a formal cause; rather, it is one (peculiarly active) constituent of the juice that gives each vegetable its specific nature. In this way the sun's old role as a differentiating light could be made consistent with the new preference for "general Laws" rather than "specifick Forms."[12]

Newton's own reason for speculating that bodies "receive much of their Activity" from light echoes his early alchemical writing; he hypothesizes that ordinary "gross bodies" may all be condensed out of the subtle matter of light—a theory "very conformable to the Course of Nature, which seems delighted with Transmutations."[13] If light constitutes all bodies, it would stand to reason that light might also be the source of the virtue or activity in each body. Newton does not base his argument on the dynamic theory that light particles are uniquely endowed with repelling force and therefore uniquely able to impart motion to grosser bodies. That is, however, the theory that eighteenth-century readers eventually extracted from his work.

They accomplished this by collapsing Newton's theory of light into his theory of the æther, which had indeed attempted to explain dynamic phenomena. Newton usually ascribed action in nature (for instance, gravitation) to "active principles" that mediated God's presence in the world. The ontological status of these active principles remained ambiguous; at times (for instance, in the General Scholium to the *Principia*) Newton hints that they are the direct expression of God's all-sustaining will; at other times they seem to be demiurgic spirits or even active substances.[14] In the queries added to the second English edition of the *Opticks*, Newton hints that an elastic æther might be the active principle accounting for gravity and for the refraction of light.[15] This æther could be composed of tiny particles separated by repulsive force—particles so small in relation to the force they embody that they hardly behave like matter at all. For instance, although the æther's elastic springiness retains the planets in their orbits, the æther itself moves freely through the pores of common matter and offers no resistance to motion. It does not produce gravitation by mechanical impulse or by pressure on the surface of bodies; it is rather a dynamic medium that pervades bodies and acts throughout their bulk.[16]

Newton's ætherial hypothesis had an extensive influence in the first half of the eighteenth century, providing a model for attempts to explain fire, light, and electricity as elastic fluids composed of tiny corpuscles separated by a force of repulsion.[17] Many of these theories simplified Newton in two characteristic ways. They streamlined his system, first, by uniting light, electricity, and "elemental fire" as modes of a single ætherial fluid. Secondly, they transformed this fluid, which Newton had used to explain gravity, into gravity's antagonist. Such a reversal might seem to involve a wholesale rejection of Newton, but it was rather a consequence of British philosophy's growing familiarity with the Newtonian world it had come to inhabit. As action-at-a-distance became a familiar assumption, philosophers felt less need to apologize for it by offering a hypothesis that would (like Newton's æther) explain remote attraction in terms of the direct contact of matter with matter. In 1734, for instance, Newton's apostle J. T. Desaguliers stated confidently that "Attraction and Repulsion seem to be settled by the great Creator as first Principles in Nature; that is, as the first of second Causes; so that we are not solicitous about their Causes, and think it enough to deduce other things from them."[18]

On the other hand, once gravitation was accepted as a first principle settled in the fabric of nature, new explanatory problems emerged. If gravitational attraction were truly universal, it would logically need to be balanced by some repelling power; the elastic, minutely particulate fluid that Newton had used to explain gravity was therefore pressed into service to oppose it. Stephen Hales states the problem clearly:

[I]f all the parts of matter were only endued with a strongly attracting power, whole nature [sic] would then immediately become one unactive cohering lump; wherefore it was absolutely necessary, in order to the actuating and enlivening this vast mass of attracting matter, that there should be every where intermixed with it a due proportion of strongly repelling elastic particles, which might enliven the whole mass, by the incessant action between them and the attracting particles . . .[19]

For Hales, the "strongly repelling elastic particles" are particles of air. In France, Gabrielle du Châtelet solved the same problem in a slightly different manner by supposing that fire was "the perpetual antagonist of gravity,"[20] tending to expand the bodies that gravity tended to contract. This hypothesis had the virtue of simplicity. The same elemental fluid that explained fire might explain electric phenomena; light and heat might be modes of its existence. Similar proposals were widely adopted in Britain around 1750.[21]

Also widely echoed was Du Châtelet's conclusion that fire's ceaseless expansive power sustains the motion and life of the universe. "[I]f matter were for an instant deprived of this spirit of life that animates it . . . everything in the Universe would be compressed, and would soon be destroyed."[22] Since the "perpetual oscillations of expansion and contraction" that animate matter are produced by the interaction of two forces, Du Châtelet might have assigned the title "spirit of life" to gravity just as consistently as to fire. But neither she nor any other eighteenth-century writer in fact chose to describe gravity in that fashion. Gravity, possessed by all portions of matter equally, perhaps lacked fire's salient variability. Just as important, the claim that life is fire came accompanied with an ancient and inexhaustible plausibility, for psychosexual reasons that Gaston Bachelard has explored. Fire is generated by rubbing, for instance; it can give birth to new fire and it digests the things it consumes into shapeless refuse.[23] The analogy between life and weight has rarely been quite as persuasive. Such considerations perhaps led writers to give repulsion (rather than attraction) credit for animating nature. At first those considerations tended likewise to promote fire in particular as the primary emblem of repelling power. Although the phenomena of light, electricity, and fire were often attributed to a single elastic fluid, the fluid was at first usually identified with fire or electricity, while light was defined as its effect, or as a mode of its existence. Light perhaps seemed poorly adapted to its new task of animating bodies: as Bachelard expresses the ancient prejudice, "light plays upon and laughs over the surface of things, but only heat *penetrates*."[24]

The Identity of Agency and Appearance

In the last third of the eighteenth century, light nevertheless displaced fire as the power that pervades living beings and sustains their motion. There were specific experimental reasons for the change. But it should be noted, first, that the prejudice Bachelard expresses so well—based on a distinction between appearance and agency—was already being overturned by empirical epistemology. John Locke's innocent premise that knowledge enters through the senses assumed increasingly radical forms after his death. Locke had used it to establish a distinction between primary qualities inseparable from matter (solidity, figure, motion), and secondary qualities (color, taste, sound) that are really only powers to produce a given sensation in an observer.[25] But as P. M. Heimann and J. E. McGuire have shown, eighteenth-century philosophers used a skeptical version of Locke's premise to overturn that very distinction. For these writers, supposedly primary qualities such as figure and solidity are really inferences that the human mind draws from other powers more directly sensed—for instance, motion and resistance. According to Newton's first rule of reasoning, philosophers should invoke no more causes of things than are both true and sufficient to explain observed phenomena. Figure and solidity could therefore be dismissed as superfluous hypotheses.[26]

This sort of reasoning led a number of natural philosophers to conclude that physical principles of causation should be reduced to a few primary sensations, which are all that the mind can ever know. Cadwallader Colden, one of the earlier writers to take this approach—and incidentally, governor of New York—discovered only two essential powers in matter: one a *vis inertiæ* that resists movement, the other a power that initiates and sustains motion, which he identified with light. Colden reasoned that "as the resisting power is the proper object or cause in our sense of feeling, so the moving power is the proper object or cause of seeing. . . . " The hesitation between "object" and "cause" concisely displays the tendency of Colden's thought, which is to equate the two categories. Light informs us of motion; light is all we know of motion; light therefore is motion. Colden erases the Lockean distinction between a body, its primary qualities (like figure), and its agency on the senses. All that remains is agency. "Every thing, that we know, is an agent, or has a power of acting: for as we know nothing of any thing but its action, and the effects of that action, the moment any thing ceases to act it must be annihilated to us: we can have no kind of idea of its existence."[27]

Colden's book was not itself influential in Britain, but it exemplifies a group of similar arguments.[28] The Scottish philosopher James Hutton used the same empirical reasoning to create the same sort of world: one where forms and qualities dissolve into pure, directly experienced agency. An account of a solid, figured body "is founded upon the resistance which is found in the body; or it is formed from the action of our mind, that is, from the exertion of our will, which we are conscious is acting in opposition to that resistance."[29] Therefore "power and action is all that, in strict reasoning, can be concluded as really subsisting externally."[30] Hutton continues to refer to "substance," but he is careful to note that he means only something unknowable that is the hypothetical seat of power and action. Instead of invoking Colden's peculiar agencies that create motion or resistance in indeterminate directions, Hutton followed well-established Newtonian tradition and defined his agencies as central forces of attraction and repulsion. The attracting force is identified with ordinary matter, which he calls "gravitating matter." The repelling force is identified with light, which he calls "the solar substance."

For skeptical empiricists[31] like Hutton and Colden, light's connection to surfaces and appearances hardly disqualified it from exerting real power.

On the contrary: appearance was agency, and that sort of agency was all the mind could ever know. To this way of thinking, light's power to produce impressions of all visible things conferred a special elegance on the hypothesis that light might produce the things themselves. The sun would in that case be the symbolic center of both knowledge and power. Experiments on the chemistry of vegetables seemed to confirm this speculation.[32] After discovering that plants could restore the air vitiated by candles or mice, Joseph Priestley reasoned that plants absorb "phlogiston," the principle of fire, from the air. By removing phlogiston from saturated air, plants make air capable of absorbing more phlogiston, and thereby render it fit again for combustion and respiration. Plants presumably store the phlogiston they imbibe within their tissues. The stored phlogiston explains why plant matter burns so well, and also why it supports animal life. When animals eat plants, the phlogiston is perhaps absorbed, converted into electric fluid by the brain, and distributed to the muscles.[33]

Priestley remarked in passing that phlogiston might have some connection to light, but his own works didn't emphasize the sun's role in vegetable chemistry. It remained for Jan Ingenhousz to connect Priestley's argument to the sun by showing that plants only absorb phlogiston, and purify air, when they are exposed to sunlight.[34] This evidence created a problem of apparent redundancy: why should plants need light (a radiant form of fire) in order to absorb the phlogiston (elemental fire) already diffused in the air? In *Mémoires physico-chymiques sur l'influence de la lumière solaire* (1782), Jean Senebier attempted to solve the problem by showing that plants were unable to absorb phlogiston in its pure form, and required a complex series of intermediate reactions. To put it briefly, the phlogiston in the air can only be separated from air through the like-to-like attraction of the phlogiston in light.[35] The details of Senebier's chemical solution were never widely accepted; the problem he wrestled with was soon defined out of existence by Lavoisier's new terminology. What did prove lastingly influential was Senebier's claim that the color and combustibility of all terrestrial objects spring from the sunlight they contain.

For Senebier, this was not only the solution of a chemical puzzle, but a discovery with the broadest philosophical implications. "It allows us to see that these corpuscles, which strike our eyes and inspire our souls with the spectacle of NATURE, contribute to the sustenance and the dissolution of portions of that spectacle by combining with them."[36] Light is at once the spectacle of nature and the power behind the spectacle.[37] It is not only the medium that carries impressions of color to the eye; it is also the chemical cause of color within bodies, since light's phlogiston produces the green of leaves, the varied tints of flowers, and perhaps the colors of minerals, animals, and artificial dyes as well. In fact light creates most of the world's sensible qualities, since light-derived phlogiston is also "the principle of savors and smells."[38] But light is not only responsible for the realm of appearance; it is at the same time a fuel, a latent and compressed animating power. "[T]hus this light, which we have blessed a hundred times along with its Divine AUTHOR, when we admire its sublime play among our cheerful perspectives . . . this light, I say, returns to illuminate our homes, and to cheat winter by banishing its frosts; it furnishes animals, by means of plants, with the phlogiston of tallow and wax; it fills wood with the warmth that reanimates us: but only Man, who corrupts everything, makes it serve in the cannon's charge to create and illuminate massacres. . . ."[39] The cannon-flash is yet another emblem of the dual role that fascinates Senebier: light is both the power creating, and the power illuminating, the spectacle of battle. If light both creates and reveals the world, it can also simultaneously reveal and destroy it.

The phlogiston chemistry on which Senebier relied was eventually replaced by Antoine Lavoisier's oxygen theory: flammable objects burn not because they lose something (phlogiston), but because they combine chemically with atmospheric oxygen. In the late 1780s, Senebier accordingly revised his understanding of vegetable physiology. He was nevertheless able to retain his fascination with latent light—remarking that wood releases, in winter, light that it seems to have stolen from the sun in summer.[40] Even Lavoisier himself supported the theory "that light combines with certain parts of vegetables," and a passage from his influential *Traité elementaire de Chimie* made an even more ambitious claim. "Organization, sensation, spontaneous motion, and all the operations of life, only exist at the surface of the earth, and in places exposed to the influence of light. Without it nature itself would be lifeless and inanimate. By means of light, the benevolence of the Deity has filled the earth with organization, sensation, and intelligence. The fable of Promotheus [sic] might perhaps be considered as giving a hint of this philosophical truth, which had even presented itself to the knowledge of the ancients."[41]

Light as Fuel for Body and Mind

In Britain, similarly sweeping claims were made for light (and for oxygen, understood as a chemical product of light) throughout the 1790s. Humphry Davy, for instance, made light the material basis not just of muscular contraction but of consciousness, and he hinted that further investigation into the chemistry of light might uncover the laws of pleasure and pain. Davy's mentor Thomas Beddoes attempted to found a new discipline of "pneumatic medicine" on the therapeutic application of oxygen. To understand why Lavoisier's observations seemed to promise such immediate therapeutic benefits, it is necessary first to understand the medical system associated with the Edinburgh physician John Brown.

Brown studied medicine at Edinburgh; after a series of public disputes with his mentor, William Cullen, he began to teach outside the medical school. He succeeded in attracting students, and soon published his theories as *Elementa medicinae* (1780). Brown's theory was radical in its simplicity: all causes of disease and all healing agents are reduced to a single power, which encompasses both the mind and the body, and explains both disease and health. Every animal is endowed at birth with a fixed amount of this vaguely defined power, which Brown calls "excitability." Stimuli—including warmth, food, exercise, labor, conversation, and opium—act upon that fund of excitability to produce "excitement" or life; in doing so, they also consume it. Sleep and rest restore part of the lost excitability, but nothing can wholly and permanently restore it. Life therefore consumes its own basis, and is itself the cause of death.

This explanation of life leads logically enough to Brown's theory of health and disease, which Thomas Beddoes, his first editor, aptly compares to a bellows-driven fire with a fixed amount of fuel.[42] If the bellows blows too weakly, the fire may sink or even go out. But if the bellows blows too forcefully, the fire will blaze until it has consumed all its fuel, and likewise go out. In Brown's theory, diseases are similarly caused by an imbalance in the expenditure of excitability. If the patient's stimuli are too weak, or if habitual stimuli are suddenly removed, not enough excitability will be transformed into excitement: this condition produces the diseases of "direct debility." Marked by the morbid accumulation of unused energy, they can easily be cured by the application of strong stimulants such as animal food, wine, and opium. If the stimuli are too strong, on the other hand, the patient will for a while enjoy a heightened degree of life: "all the senses are acute, the motions both involuntary and voluntary are vigorous, there is an acuteness of genius, great sensibility, and a tendency to passion and emotion."[43] But such a life quickly burns itself out; and since excitability can never be restored, the diseases of "indirect debility" that result are much more difficult to cure than the simpler diseases of direct debility. Fortunately, they are also less common.

The evidence for Brown's influence on medical practice is ambiguous. Certainly more doctors began to prescribe wine and opium as stimulants; it remains unclear how far this was due to Brown's influence, and how far it was due to patients' understandable preference for that sort of medicine.[44] Brown's importance for late-eighteenth-century medical theory, on the other hand, is indubitable. His radically simplified account of disease seemed to show how chemical theories of life could be put to immediate use. If all diseases were merely the consequences of an excess or deficiency of vital action, then understanding the chemistry of vital action would allow doctors to eliminate disease. In 1790, the Göttingen physician Christoph Girtanner published a two-part article that grafted Brown's medical theory onto Lavoisier's theory of respiration.[45] Girtanner used Albrecht Haller's term "irritability" rather than Brown's "excitability," and he followed Haller in identifying irritability specifically as the power of muscular contraction. But he applied the concept in a Brunonian way by defining health as a balance between the quantity of irritability in the living system and the quantity of stimulus applied to it. He also divided illnesses according to their origin in an excess or a deficiency of stimuli. Girtanner's irritability is not fixed at birth, as Brown had claimed; it is nothing other than oxygen, supplied continually by the lungs. Health therefore consists in a dynamic balance between the supply of, and the demand for, oxygen.

In England, Thomas Beddoes had been independently struggling to apply the new gas chemistry within a Brunonian medical framework; joining forces, Girtanner and Beddoes began to cite each other and translate each other's works. Beddoes, especially, sought to translate chemical theories of excitability into therapeutic recommendations. Scurvy, Beddoes thought, was caused by a deficiency of oxygen. Since "oxygen" was named for its presence in acids, the recognized utility of acid fruit in curing scurvy seemed to confirm this hypothesis: "no substances are better calculated than acids at least, to impart oxygene to the system." This hypothesis might also explain "the observation which has so frequently been made at sea, that the scurvy makes its appearance after a storm, when the seamen, having undergone violent exercise, have expended a great part of the oxygene of the solids." Since old age itself could be a creeping deficiency of excitability or oxygen, Beddoes speculated that it would be possible to restore excitability chemically, and thereby "to protract the period of youth and vigor indefinitely."[46] In 1798, he established a Pneumatic Institute in Bristol to investigate the medical applications of oxygen and other "factitious airs."

Although Girtanner and Beddoes certainly knew that oxygen was produced by plants exposed to sunlight, neither physician explicitly emphasized the link between sunlight and human respiration. But a number of other writers did emphasize that link, by suggesting that atmospheric oxygen contained sunlight, and that sunlight was responsible for oxygen's vivifying power. In a paper published in 1786, Claude Louis Berthollet suggested that gaseous oxygen was actually a compound of oxygen and light. This seemed to explain the fact that certain acids released oxygen gas only when exposed to light; it might also imply that plants release oxygen because light fosters the decomposition of water in their tissues. "Since light has such a great influence on vegetation," Berthollet continues, "might it not also have an influence on certain animal functions? Are not a number of effects attributed to air, to heat, or to other causes really due to light? Perhaps this is responsible, for instance, for the disadvantages that have been noted in the habit of substituting sleep in the day for sleep at night. Thus the beauty of day, the purity of the sky not only lend nature charm in our eyes, but vivify her."[47]

Berthollet's theory seems to have influenced Lavoisier's thought about light and oxygen in the late 1780s. In his *Traité elementaire de Chimie* (already quoted), Lavoisier described the myth of Prometheus as an allegory of life's dependence on light. In the "Premier Mémoire sur la Respiration," cowritten with Armand Seguin, he describes respiration in similar terms.

> This fire stolen from the sky, this torch of Prometheus, is not only an ingenious poetic conception; it is the faithful image of the operations of nature, at least for those animals that breathe: one might thus say, with the ancients, that the torch of life is lit at the instant when the infant breathes for the first time, and is not extinguished till death.[48]

The connection Lavoisier established between respiration, life, and "fire stolen from the sky" was not lost on the young Humphry Davy, who independently formed a chemical theory similar to Berthollet's. Davy used Lavoisier's remark about light from the *Traité élémentaire* as the epigraph for his own "Essay on Heat, Light, and the Combinations of Light." The essay was published, as it happens, under Thomas Beddoes's editorship; by 1799, Davy was working as Beddoes's laboratory superintendent at the Pneumatic Institute. But the experiments described had been carried out the preceding year, and reflect Lavoisier's influence more than that of Beddoes.

Like Berthollet, Davy asserts that the gas named oxygen is not an element at all, but rather a compound formed from an unknown base—the real oxygen—and light. Since this compound contains latent light, Davy assigns it a new name: "phosoxygen." The production of phosoxygen shows that plants are "hydrogen attractors"; the combination of their affinity for hydrogen and light's affinity for oxygen is sufficient to decompose water (*DW* 2:44–47). The process creates a vast reservoir of phosoxygen in the atmosphere, which animals inhale and absorb in their blood. Somewhere within the body, phosoxygen decomposes and the oxygen itself is exhaled again in carbonic acid. What happens to the light? Davy reasons that the decomposition of phosoxygen takes place in the brain, and that "light is attracted or secreted from the blood by the brain, and perpetually conveyed by the brain to the nerves. . . . On this supposition, sensations and ideas will be motions of the nervous ether or light exciting the medullary substance of the nerves and brain into sensitive action" (*DW* 2:82). Animals breathe, in short, to obtain light for sensation, motion, and thought.

Davy is thus able to argue that light provides the material basis of "perceptive"—which is to say, conscious—life.

> We may consider the sun and the fixed stars, the suns of other worlds, as immense reservoirs of light destined by the great ORGANISER to diffuse over the universe organization and animation. And thus will the laws of gravitation, as well as the chemical laws, be considered as subservient to one grand end, PERCEPTION. (*DW* 2:85)

One can trace in Davy a Senebier-like fascination with light's dual role as agency and appearance, but the spectacle that light at once reveals and constitutes has shifted its location to the human mind. The medium that conveys "the most numerous and most pleasurable of our perceptions" through the air turns out to be the same power that constitutes "perception" and "pleasure" inside the human brain (*DW* 2:25). Davy took a particular interest in the connection between light and pleasure, because he hoped that research into the laws of perceptive life would allow chemists "to destroy our pains and to increase our pleasures," thereby releasing chemistry from its merely instrumental role, and making it a direct contributor to human happiness: "the most sublime and important of all sciences" (*DW* 2:85–86).

Davy's essay on light was unusually ambitious and speculative, but it was by no means unique in its period. Beyond the works by Berthollet, Lavoisier, Girtanner, and Beddoes already cited, one might examine John Thelwall's *Essay Towards a Definition of Animal Vitality* (1793), which—Nicholas Roe has suggested—may have helped to shape the premise of "One Life" in the early works of Wordsworth and Coleridge. Although Thelwall describes the stimulus powering life as an "electrical fluid" rather than light, he similarly understands that fluid as a component of the atmosphere, and assumes that it is absorbed in respiration.[49] Light and electric fluid were frequently yoked together in theories of this sort. The American physician Benjamin Waterhouse, for instance, proposed in 1790 that animals breathe to obtain an electrico-solar fluid diffused throughout the atmosphere.[50] In Philadelphia, Joseph Trent published an *Inquiry into the Effects of Light in Respiration* (1800) that closely parallels Davy's conclusion: light is a constituent part both of oxygen gas and of human blood. Two years later a Brunonian physician in New York described respiration as "natural insolation," and recommended medical applications of electricity ("artificial insolation") to cure those patients whose insufficient supply of solar substance had produced debility.[51]

The Difference between Vital Energy and Toil

Girtanner, Beddoes, Davy, Waterhouse, and Trent all sought to explain life in terms of an expendable vital principle contained in the air: a fuel, as it were, that human beings inhale. Beddoes linked that vital principle to physical labor; in texts that will be considered in chapters 2 and 3, Davy linked it to intellectual work. But none of these writers made any attempt to measure a ratio of equivalence between oxygen and work. This may seem a surprising omission. But then, most of these writers were physicians. Insofar as they were concerned with their patients' labor at all, they saw it as a therapeutic tool. Labor was one of many things they could urge patients to forego or to take in greater quantity—a free rather than a dependent variable. For this reason, John Brown categorizes labor as a stimulant, along with animal food, opium, and conversation. Thomas Beddoes thought that excessive labor might be a cause of scurvy, and therefore advised scorbutic patients not to exhaust themselves. But determining the quantity of labor that could be produced by a fixed quantity of oxygen was not part of his professional concern.

Antoine Lavoisier was not a physician, but a senior administrator serving the French state in a variety of capacities: reforming agriculture, directing the production of gunpowder, and calculating the total national revenue of France. He did, accordingly, make the equation between labor and oxygen part of his professional concern. In 1789, Lavoisier and Armand Seguin published the "Premier mémoire sur la réspiration des animaux," which goes farther in attempting such a correlation than anything else written in the eighteenth century. Lavoisier and Seguin demonstrate that human consumption of oxygen increases with exercise—in fact, they postulate that an experimental subject's consumption of oxygen will be in a roughly geo- metrical proportion to exertion, which they measure as weight lifted through a given height.[52]

> This sort of observation allows one to compare uses of force between which there might seem to be no relation. One could discover, for instance, what weight in pounds equals the effort of a man who reads a discourse, or of a musician who plays an instrument. One could even measure what there is of a mechanical nature in the work of a philosopher who reflects, of a man of letters who writes, or of a musician who composes. These effects, though considered as purely moral, have something bodily and material in them that allows them to be compared, in that respect, with the things done by a manual laborer. It is thus not without a certain justice that the French lan- guage has confounded, under the common denomination of *work*, the efforts of the spirit with those of the body, the work of the study and the work of the shop.[53]

This insight is strikingly unusual for the age. Not that Lavoisier was alone in claiming that mental tasks exhaust the body just as physical tasks do. Brunonian physicians made the same assumption. Nor does quantification itself set Lavoisier's work apart: Brunonian physicians had created numbered scales of excitability and sought to use them as a therapeutic guide. What is unusual about the paper by Lavoisier and Seguin is that it proposes a dimension of measurement very different from excitability. Instead of attempting to measure a power understood as an attribute of individuals, Lavoisier sought to establish a correlation between tasks accomplished and quantities of oxygen consumed.

Lavoisier's emphasis on the correlation of tasks and volumes of oxygen is interesting because it sharply contrasts with the usual eighteenth-century approach to the measurement of work capacity. The first impulse of writers who dealt with this problem was to measure the power possessed by an agent. How much can a single laborer lift? How many men equal the power of a horse? How many horses equal the power of a steam engine of a given size? Engineers knew that waterpower could also be measured, not as the power of a wheel, but as the total quantity of water available at a given height. But the importance of this measurement was not widely appreciated by writers in other fields. Not until the 1820s did a consensus emerge that work capacity could be measured in terms of the total effect produced, without specifying the number of agents involved or the time elapsed.

In this respect, the "Premier mémoire sur la réspiration" is typical of a heuristic that Lavoisier applied to chemical, financial, and administrative domains. In each of these fields, he attempted to solve problems by creating a closed system, so that it became possible to establish a reliable correlation between the quantity of one substance consumed, and the quantity of another produced. One might, for instance, calculate the total quantity of alcohol and tobacco being consumed in Paris, and compare it to the quantity of those goods entering the city, as registered by customs officials. If the two figures do not match, the difference between them must represent a certain quantity being smuggled; it then becomes necessary to build a wall around Paris, in order to establish a more reliably closed system. Such a wall was in fact built at Lavoisier's suggestion.[54] The same technique might be applied to the design of an insulated calorimeter. In the case of the "Premier mémoire," Seguin himself served as the closed system: he pumped a foot treadle while wearing a specially-designed mask so that his consumption of oxygen could be measured and correlated with his production of dynamic effects.

Lavoisier's insistent attempts to establish a reliably closed relation between production and consumption are almost certainly related to his work as a political economist and administrator. Bernadette Bensaude-Vincent has argued, in particular, that the gravimetric balance and the balance sheet are intimately linked in Lavoisier's thought, which she describes as "a science of balance sheets."[55] Lavoisier's administrative responsibilities may also do something to explain his interest in the subject of mechanical

Figure 1.1 "Experiment on the respiration of a man doing work," print by Marie-Anne Paulze Lavoisier. Division of Rare and Manuscript Collections, Cornell University Library. Page 42.

work, which was not an ordinary research subject for late-eighteenth-century chemists. The "Premier mémoire sur la réspiration" of Lavoisier and Seguin thus foreshadows on several levels the close interaction between political economy and physical science that would eventually create a systematic philosophy of production in the 1820s.

But Lavoisier's writings actually did little to create that discourse. The "Premier mémoire" is a precursor only in retrospect, because this aspect of his research was not received actively by other investigators. Why didn't other chemists pursue the same research program? Humphry Davy, for instance, was familiar with Lavoisier's writing, and actively investigated the physiological effects of "factitious airs" for several years. But it never occurred to him to attempt an experiment like Lavoisier's. In his research on nitrous oxide, Davy did loosely compare quantities of gas inhaled with amounts of motion produced—but both Davy and his experimental subjects stressed that the motion involved was spontaneous. Davy didn't sit his friends Beddoes and Coleridge down at a foot treadle to measure the amount of controlled mechanical work they could perform after breathing nitrous oxide. In fact the experimental point of their motion lay precisely in its accompanying freedom from control. Davy called the effect "an irresistable propensity to action"; for one of his subjects, Dr. Kinglake, it was "a sense of additional freedom and power (call it energy if you please)" (*DW* 3:272, 3:298). These descriptions seemed to show that nitrous oxide was the chemical source of human energy that Davy had for several years been seeking: a principle that produced at the same time pleasure, heightened perception, and spontaneity.

Part of Davy's experimental approach may be due to the fact that he chose to work on nitrous oxide, which admittedly does not lend itself especially well to sober correlations. But his approach was also consistent with his previous assumptions about the nature of human energy. All along he had expected it to be something like a power of spontaneity or initiative. As chapter 2 argues, this assumption about vital power was related to a social division between two kinds of work: "mechanic toil," a form of constraint, and "spontaneous energy," which flows from within and produces freely without toil. At bottom this division expressed the middle classes' desire to identify their own work with the propertied independence of their social betters, rather than with the hired drudgery of their inferiors. From this perspective, what is unique about Lavoisier is his willingness to measure "the work of the study" and "the work of the shop" in units that could also be used for the toil of the laborer, thereby reducing the work of all classes to the same mechanical dimension. In late-eighteenth-century Britain, by contrast, connections between natural force and human work were shaped so as to exclude mechanic toil from the calculation.

CHAPTER 2

ENERGY AND THE AUTONOMY OF
MIDDLE-CLASS WORK

Until the eighteenth century, *energy* was a learned term, common only in texts of metaphysics and rhetoric. After 1750, it rapidly became a buzzword. The frequency of *energy* and *energetic* in the written record tripled between 1750 and 1800.[1] (The frequency of the French cognate, *énergie*, quadrupled over the same period.[2]) The word began to appear in novels and in personal letters; in the last quarter of the century, energy-talk was so much in vogue that it became an object of parody on both sides of the English Channel.[3] This chapter argues that the new popularity of *energy* and its cognates was bound up with a change in the expression of the work ethic. Instead of describing work as a conscious product of the moral will, writers began to compare the energy of the worker to the spontaneous energies of nature.

The twenty-first century still uses this word to compare human work and natural force. But contemporary use of the term has been shaped by nineteenth-century physics and political economy; energy is now a quantity of something that can be stored and spent. In the eighteenth century the word had no mathematical definition, and didn't necessarily designate a quantity at all. Natural forces like heat and electricity didn't possess measurable amounts of energy; rather, they were energies. The word was not restricted to natural philosophy; language and sentiment were energies as well. Any action could, in principle, be described as an energy.

What set *energy* apart from its near synonyms was a looseness of definition that could be interpreted as spontaneity. Words like *power* and *force* tended to imply a transitive relation between their possessor and the world: a power over some object, or a force to accomplish some specified effect. Much the same thing was true of social synonyms like *diligence, passion,* or *industry* that implied forms of activity determined by particular duties, desires, or goals. *Energy* named an action without implying any object, and could be used to describe forms of agency that were spontaneous and complete in themselves.[4] As it happened, writers in more than one eighteenth-century discipline were in need of precisely such a term. Moral philosophers were searching for a way to describe the emotions, not as "passions" or yearnings for a specified object, but as morally beneficent forms of mental exercise. Natural philosophers needed a name for matter's inherent activity. Political writers were looking for a phrase that would describe middle-class work as a form of autonomy, rather than compulsion. All three groups seized on the same word.

The parallel among these three developments is suspicious. Theories of matter's independence from spirit may owe their growing popularity, as Steven Shapin has argued, to commoners' intuition of their own potential independence from church and king.[5] In this chapter, I explore a slightly less ambitious hypothesis that can be established with greater certainty. Republican political feeling may or may not have motivated the development of eighteenth-century matter theory, but it is clear that the parallel between the two discourses had become conscious and overtly political by the end of the century. British novelists and philosophers of the 1790s compare the autonomous energies of nature to the autonomous energies of the human mind, and do so explicitly in the service of middle-class political interests.

The Energy of the Passions

Energy descends to modern European languages mainly through Aristotle. In the *Metaphysics*, it is one of three forms of movement: *dynamis* (potential), *energeia* (activity or actuality), and *entelecheia* (completion). When a bowstring is taut, the bowshot exists only as *dynamis*; when it is released, the motion of the arrow is *energeia*; and the presence of the arrow in the target is the *entelecheia* of the shot.[6] A separate rhetorical sense of the word emerged from a misinterpretation of Aristotle's *Rhetoric*.[7] *Energeia* means activity there, just as it does in the *Metaphysics*: Aristotle advises speakers to describe activity, or to describe things as if they were active, in order to create a vivid effect. But Aristotle's medieval interpreters blurred the distinction between the activity described and the vivid effect produced. The resulting confusion is too complex to detail fully here.[8] Suffice it to say that by the late sixteenth century the rhetorical meaning of *energy* had been worn down to "force or vigor of language." Sidney's *Defence of Poesie*, for instance, briefly discusses the "forciblenesse or *Energia* (as the Greeks call it) of the writer."[9]

In the early eighteenth century, the English word *energy* could thus be used in a variety of different ways. At times, it was close to Aristotle's own meaning: simply a synonym for "activity." At other times, the word described an especially forceful and vigorous quality of language—or more broadly, a forceful and vigorous quality belonging to anything. Around the middle of the century, the word began to develop a specialized application that drew on both of these earlier meanings. Writers began to use *energy* to describe sentiments and affections as if they were autonomous agencies. In his *Covent-Garden Journal*, for instance, Henry Fielding writes that the pleasure of the passions lies in "the Energies themselves."[10] Fielding could just as well have written that the pleasure of the passions lies in "the exercise of them," except that there is something a bit oxymoronic about exercising a passion. A passion can hardly act or be exerted. *Energy* provided a way of avoiding this oxymoron while referring to the affections as active powers— which is how sentimental moral philosophy was coming to conceive of them. A few years earlier, in *Tom Jones*, Fielding had written of "all those strong energies of a good mind, which fill the moistened eyes with tears, the glowing cheeks with blood, and swell the heart with tides of grief, joy, and benevolence."[11] Here, again, *energy* allows Fielding to describe feeling not as a disturbance that the soul undergoes or suffers, but as a mode of benevolent action—one, incidentally, that constitutes its own reward.

The French word *énergie* specialized in a similar way around the same time. Denis Diderot's 1745 translation of Anthony Ashley Shaftesbury's "Inquiry Concerning Virtue or Merit" often inserts the word gratuitously into Shaftesbury's text, in order to reframe an intensity of passion as an intensity of action. A passion is allowed "full scope" in Shaftesbury, for instance, but acts "dans toute son énergie" in Diderot.[12] Not long afterward, Jean-Jacques Rousseau writes in *La nouvelle Héloïse* (1761) of "cette énergie de sentiments qui caractérise les âmes nobles."[13] Rousseau uses the word with a lingering consciousness of its rhetorical meaning; *énergie* is a kind of vehemence or animation that lifts the soul above mundane existence.

Texts like *La nouvelle Héloïse* probably did as much as Fielding to shape Ann Radcliffe's interest in the energy of passion. Describing Montoni, the chief villain of her *Mysteries of Udolpho*, Radcliffe writes that he was "little susceptible of light pleasures," but "delighted in the energies of the passions." "Energies of the passions" is plural here in the way "waters of the Nile" is plural. Radcliffe is not discriminating between separate energies; the plurality of the word merely allows her to suggest contending forces that baffle each other like winds in a storm.

> He delighted in the energies of the passions; the difficulties and tempests of life, which wreck the happiness of others, roused and strengthened all the powers of his mind, and afforded him the highest enjoyments, of which his nature was capable. Without some object of strong interest, life was to him

little more than a sleep; and, when pursuits of real interest failed, he substituted artificial ones, till habit changed their nature, and they ceased to be unreal.[14]

The sublime villain's delight in opposition is primarily a delight in the exercise of his own powers of mind—so much so that it doesn't especially matter whether these energies are exercised against real obstacles, or against obstacles of his own invention. This is an extreme example of *energy*'s general tendency to connote autonomy. Radcliffe is slightly less willing to grant this extreme autonomy—which verges on solipsism—to her heroines. "Hers was a silent anguish . . . not the wild energy of passion, inflaming imagination, bearing down the barriers of reason, and living in a world of its own."[15]

Energy and Natural Philosophy

The history of matter theory in the eighteenth century has been studied in depth, and needs only brief retelling here.[16] At the end of the seventeenth century there was widespread agreement (though not absolute consensus) that matter was passive, and moved only through the intervention of some external force or spirit. John Locke, for instance, wrote "A body at rest affords us no idea of any active power to move; and when it is set in motion itself, that motion is rather a passion than an action in it."[17] "Sluggishness," a lack of active initiative, was often offered as a defining characteristic of matter.[18] Isaac Newton agreed, although there were instabilities in Newtonian philosophy that opened up the possibility of other opinions: Newton had little evidence, for instance, to support his assumption that the force of gravity was imposed on matter rather than inherent to it.[19] By 1790, the situation was reversed. Matter had acquired active power; it was indeed commonly defined as active power, and natural philosophers relied on a general heuristic that in nature "there is no reason to look for any thing inactive or absolutely passive."[20] An understanding of matter's essential activity, in the words of one writer, "removes the difficulty which has ever attended the question of the origin of motion, by showing motion to be the original form of being."[21]

Philosophers did not, however, agree about the nature and cause of matter's activity. Some believed, with Joseph Priestley and Roger Boscovich, that all particles of matter could be described as mathematical points without extension, obeying a single force of attraction and repulsion that varied with the distance between each pair of points. Others preferred to distinguish between ordinary, sluggish matter, and a special active fluid that they conceived to pervade ordinary matter as water pervades a sponge. Others, especially physicians, stressed the importance of internal organization—particles of matter become active only when they take the form of, say, contractile fibers. For writers of natural philosophy, the advantage of the word *energy* was that it could bracket this whole dispute. In metaphysics, *energy* had been a mere synonym for "activity" or "operation." To say that matter "displays energy" was simply to say that it was active, while declining to offer a hypothesis about the source of action.

But writers who declined to offer a hypothesis were themselves making a choice. To bracket the debate about the microstructure of matter was in practice a way of redirecting readers' attention toward agencies like electricity and light, agencies that (though perhaps material) seemed to obey no known mechanical laws. In 1772, for instance, the Edinburgh-trained American physician James MacLurg set out to resolve the conflict between two schools of medical thinking: the iatromechanists—who attempted to explain life in purely mechanical terms—and the spiritualist followers of Georg Stahl. He did so by insisting on the primacy of an intermediate category, energy. This category was intended to supersede both mechanics and the soul as an explanatory principle.

> The animal machine, though it is connected with the other parts of Nature, is distinguished from them; and, while it acknowledges the influence of the common laws of matter, is governed by a principle peculiar to life. Nor do I mean, by the vital principle, the soul of the Stahlians; but that energy which discovers itself particularly in muscular contraction; and is evidently different from elasticity, or any other known power in nature.[22]

Energy could conceivably mean no more than activity here. The statement that the energy of life "discovers itself" in muscular contraction might imply that the energy is nothing but muscular contraction. That would be a fairly modest use of the word: to describe a form of activity in nature, not to assign a cause. But MacLurg also states that life "is governed by" this energy, and he goes on to show that he understands energy as a power or faculty latent in the living organism—in particular, in the nervous system. Although he isn't confident that this vital energy is electricity, he admits that the example of electricity has been an important influence, liberating medicine from purely mechanical models of causality.

> For though it is by no means demonstrated, that the nervous power is the same with that which occasions the phenomena of electricity . . . yet the contemplation of such active and subtile energies has enlarged our views, and drawn the attention from those principles of mechanics and chymistry, to which it was before too slavishly attached (lxii).

Here *energy* is clearly a "power" that "occasions . . . phenomena" rather than a name for the phenomena themselves. It is, in fact, a general term for the category of nonmechanical explanatory principles that MacLurg wants to emphasize: in some sense material, yet different from common matter in being "active and subtile." Note also that energies liberally "enlarge our views," whereas mechanical and chemical causation seem to partake (by association) of the slavishness they inspire in researchers. MacLurg's distaste for "mechanical" causation is apparently colored by the word's double meaning: it connotes not just machine-like pushing and pulling, but the class of dependent workers who earn their livings with their hands.

Electricity provided one important model for the new class of "active and subtle energies." From the 1740s on, the power of electricity had been impressively demonstrated with the Leyden jar, and contemporary British observers testified that those demonstrations shaped their own ideas about the activity of matter.[23] But the fundamental advantage of *energy* had less to do with electricity in particular than with the word's ability to finesse the distinction between describing an observed action, and reifying that action as an underlying cause. Alexander Wilson writes in *Philosophical Transactions*, for instance, "Nature unquestionably abounds with numberless unthought-of energies, and modes of working."[24] The repetition of synonyms is making a surreptitious claim. To say that nature abounds with "modes of working" is simply to say that it abounds with phenomena. No one could deny that, surely, or find it theologically dangerous. "Numberless . . . energies" is ostensibly a mere synonym for "modes of working," but the noun *energy* begins to solidify each mode of working into its own self-sufficient cause; in doing so, the word sidles quietly toward the less ortho-dox claim that nature contains all the causes of its own activity.

The Political Provenance of *Energy* in the 1790s

In short, the word *energy* was associated with autonomy in several different discourses. Radcliffe referred to "the wild energy of passion, bearing down the barriers of reason and living in a world of its own."[25] Natural philoso-phers used the term to describe self-sufficient agency in nature, as distin-guished from slavish mechanical causation. In the poems of William Hayley—a Whig poet best known now as a friend of Anna Seward and patron of William Blake—there is even a noticeable tic of rhyme whereby the appearance of "*energy*" seems to call forth the rhyme "mind / unconfin'd." In an "Essay on Epic Poetry" he refers to the Orient as

> those happier lands
> Where every vital energy expands;

Where Thought, the golden harvest of the mind,
Springs into rich luxuriance, unconfin'd.

Later in the same poem, the association of energy with mental freedom is
expressed again, and with the same rhymes. The Enthusiast

seeks to gain, by no mean aims confin'd
Freedom of thought and energy of mind;
To raise his spirit, with aetherial fire,
Above each little want and low desire.[26]

"Energy of mind" serves here as a middle term, where political "freedom of
thought" can merge with an "aetherial fire" understood to animate the
brain itself. The word functioned similarly when Edmund Burke remarked,
in 1777, that "It [liberty] is not only a private blessing of the first order, but
the vital spring and energy of the state itself, which has just so much life and
vigor as there is liberty in it."[27]

Everything I have observed so far would seem to indicate that the
connection between energy and autonomy was common intellectual property. The word is more common in the writing of Whigs like Burke and
Hayley than it is in, say, Samuel Johnson, but there is little to suggest that it
was perceived as a party slogan in Britain before 1790.[28] The word did
become a slogan, however, in the last decade of the eighteenth century. Its
political connotations developed rapidly and were felt most strongly by
younger writers. Writers born before 1740—Edmund Burke (1729–1797)
and Joseph Priestley (1733–1804), for instance—never acknowledged that
energy was the property of a single party. As the political controversies of the
1790s unfolded, however, younger British radicals began to identify themselves as the "energetic" part of the British nation. By the end of the decade,
the word energy had become a recognizable tic in radical writing, and was
isolated by younger conservative writers as an object of satire. Elizabeth
Hamilton (1758–1816), among others, pokes fun at energies in order to
foreground the sentimentality and selfishness of radical philosophy.

One way to assess the political provenance of energy in the 1790s is to list
British writers from the period who made noticeably heavy use of the
word. Such a list would include William Blake (1757–1827), Samuel
Taylor Coleridge (1772–1834), Henry Fuseli (1741–1825), William
Godwin (1756–1836), Richard Payne Knight (1750–1824), Ann Radcliffe
(1764–1823), Thomas Taylor (1758–1835), John Thelwall (1764–1834), and
Mary Wollstonecraft (1759–1797). It amounts to a list of writers with radical associations who had not yet reached their fiftieth year when the decade
of the 1790s opened. The political views of Godwin and Wollstonecraft are

best known, but Thomas Taylor and Henry Fuseli belonged to the same social circle. John Thelwall was the period's best-known radical lecturer, and Coleridge sympathized (at first) with Thelwall. Neither Radcliffe nor Knight was explicitly radical in politics, but both were tarred with the name "Jacobin" at one time or another.

Why, then, was this word so central to the vocabulary of literary radicals in the 1790s? I will approach this question by looking at three kinds of uses radical writers found for the concept of energy. First, *energy* was used to translate an individualism based on the dissenting Protestant conscience into sentimental terms. Second, it implied a self-powering activity that could be used to explain human perfectibility. Last, and most importantly, middle-class work was described as an expression of spontaneous *energy* in order to imply that it was, not a form of need or constraint, but a foundation for independence and political virtue.

It makes sense to begin by examining the role *energy* plays in William Godwin's *Enquiry Concerning Political Justice* (1793), a text whose centrality to the radicalism of the 1790s has been widely acknowledged. The political system Godwin proposes in this volume is founded above all on the free exercise of private judgment, which is, according to Godwin, the only way to determine what is just. Although Godwin defines justice in depersonalized terms, as the good of the whole, he insists unwaveringly that abstract justice can only be perceived and enacted by individuals; the free exercise of private judgment—"a doctrine . . . unspeakably beautiful"—is so much the foundation of virtue that a deed otherwise good can even lose its moral value if it is performed under coercion (*GW* 3:76, 3:73). The unspeakable beauty Godwin perceives in private judgment descends, in part, from a long tradition of dissenting Protestant emphasis on the sacredness of the individual conscience. (Godwin, like several other radical intellectuals, began his career by training to become a Dissenting minister.) But Godwin no longer names the principle that binds private judgment "conscience." Instead, he writes, "every man is bound *by the exertion of his faculties* in the discovery of right, and to the carrying into effect all the right with which he is acquainted" (*GW* 3:76, my italics). Godwin thus founds his faith in private judgment on a faith in the natural activity, or (as he more frequently puts it) the "energy" of the mind.

The term *energy* became useful for Godwin because of the implication of autonomy it brought with it from the discourses of sentiment and of natural philosophy. By referring to *energy* Godwin could describe mental power while assuming implicitly that such power must, by its own nature, be autonomous and spontaneous. The assumption plays a crucial role, for instance, in Godwin's argument against deliberative assemblies. Since deliberative action is not autonomous, it cannot, by definition, possess energy;

"men who act under the name of society, are deprived of that activity and energy which may belong to them in their individual characters" (*GW* 3:308). Godwin insists on describing judgment in terms of *energy* because this provides a way of subtly assuming that it must spring wholly from the activity of a single mind.

Godwin's concept of "intellectual energy" also updates the Dissenting conscience for a sentimental age. In emphasizing the importance of the mind's natural ardor, activity, or energy, Godwin often merges his idea of (rational) private judgment with the sentimental doctrine that virtue is measured by the intensity of passion in a human breast. In a chapter entitled "Of the Connexion Between Understanding and Virtue," for instance, Godwin argues that it is impossible for someone with a "weak understanding" to be truly virtuous; for although such a person may serve good ends, they cannot distinctly conceive and deeply feel them (*GW* 3:143). If weak-minded people are never eminently virtuous, it is conversely difficult for someone of "great intellectual energy" not to be. "Can great intellectual energy exist without a strong sense of justice?" Godwin asks rhetorically; the answer he expects is "No" (*GW* 3:146). It is for this reason that even Milton's Satan ends up appearing to most readers "a being of considerable virtue," although "it must be admitted that his energies centered too much on personal regards." Godwin generalizes from this example in the following terms:

> Upon the whole it appears, that great talents are great energies, and that great energies cannot but flow from a powerful sense of fitness and justice. A man of uncommon genius is a man of high passions and lofty design; and our passions will be found in the last analysis to have their surest foundation in a sentiment of justice. (*GW* 3:146)

Godwin is concerned to say that talents are energies because the term allows him to argue that intellectual talents necessarily spring from an innate and passionate moral sense. Here, again, Godwin relies on *energy*'s ability to connote a kind of activity that contains its cause within itself. Mere talents might perhaps be external acquisitions, but energies implicitly spring from the mind's most deeply rooted dispositions. In a phrase like "great energies," Godwin's rationalism, his individualism, and his sentimental emphasis on the passionate soul, all converge.

The idea that "mind is, in its own nature, essentially active" also undergirds the Godwinian faith in human perfectibility and progress. Godwin commonly ascribes energy to intellectual principles in order to represent their natural capacity for action. Truth, for instance, is energetic—that is to say, self-acting.

> Truth for a long time spreads itself unobserved. . . . But it goes on to be studied and illustrated. . . . The number of those by whom it is embraced is gradually enlarged. If it have relation to their practical interests, if it show them that they may be a thousand times more happy and free than at present, it is impossible that in its perpetual increase of evidence and energy, it should not at last break the bounds of speculation, and become an animating principle of action. (*GW* 3:319)

This passage tells a story about the metamorphosis of truth into an autonomous agent. Eventually, "in its perpetual increase of evidence and energy," truth ceases to be merely propositional and emerges as "an animating principle of action." Godwin uses similar language elsewhere to explain why truth compels its own propagation. "If the truths I have to convey be of an energetic and impressive nature . . . it will be strange if they do not at the outset excite curiosity in him to whom they are addressed" (*GW* 3:397). The same language can be applied to virtue. Virtue, like truth, "perpetually renovates itself" and "propagate[s] itself"; it thus follows that "all that is to be asked on the part of government in behalf of morality and virtue is a clear stage on which for them to exert their own energies" (*GW* 3:134).

The energy of the human mind as a whole could also be offered as a foundation for improvement and human perfectibility—an argument that is perhaps best exemplified by a passage, not from Godwin, but from a writer who studied him closely, Humphry Davy. The following passage dates from about 1799, when the author, not yet a famous chemist, still espoused radical politics. It is part of a fragmentary plan, never published, for improving the human race through pneumatic medicine.

> From experience . . . and . . . from the nature of the human mind, I believe that Man is capable of improvement, & this hope is a perpetual source of pleasure to me. Mortality does not make man the slave of a certain destiny. He is an active energetic being born endowed with great powers & capable of applying them; his miseries in a great measure depend upon ["his indol" crossed out] himself.[29]

In this manifesto energy has two different but parallel rhetorical functions. It is, first, the characteristic of the human mind that proves man capable of improvement. Man, though mortal, is "an active energetic being, . . . endowed with great powers" that allow him to reshape his destiny. On a secondary level, those who believe this claim are themselves marked by their energies.

> One thing however is certain. The energies of a certain portion of the human species are awakened, & there exist those who think that man was neither

born to suffer eternally moral & physical evil.—There exist those whom rational scepticism leads alone to exertion—who are confident of nothing but who hope for much. They are lighted by the lamp of true philosophy. . . . [30]

The energies that are fundamental to the human mind are thus also peculiarly characteristic of radical philosophers, that "portion of the human species" whose energies have been "awakened" to exertion.

This begins to suggest that the ideal of energy may have appealed to middle-class intellectuals, not merely through its utility in radical arguments, but more fundamentally as a representation of their own class. Although late-eighteenth-century radicalism in Britain had continuities with earlier "republican" oppositional traditions, it diverged from them in becoming, as Isaac Kramnick has argued, a specifically "bourgeois radicalism"—that is to say, "a self-conscious glorification of the mission of the middle class in English society."[31]

> The demands of the reformers that the suffrage be extended to industrial and commercial wealth, that new manufacturing centers . . . be granted parliamentary representation, that expensive aristocratic institutions be streamlined or eliminated, that Dissenters be free to serve as municipal and governmental officials, all boil down to the bourgeois demand for the opening of careers to the talented. . . . Given a freely competitive environment, these talents would move them to the top, a victory for virtue as well as for merit.[32]

Kramnick's thesis here requires some qualification. In particular, his use of the word "bourgeois" is open to question. Critics of Kramnick's interpretation—such as J. G. A. Pocock—have rightly noted that there is little evidence of a homogenous, self-conscious, and specifically urban "bourgeoisie" in late-eighteenth-century England.[33] The middle ranks of life were defined more loosely than in France, and were certainly not limited to artisans and shopkeepers: the term could expand strategically to include farmers, or the clergy, or even (in some cases) the lower ranks of the landed gentry. If, on the other hand, there is something Whiggish about Kramnick's tendency to use a specific class fraction—manufacturers, artisans, and Dissenters—to model the interests of a larger and more diffuse British "middle class," this may actually be a Whiggishness that captures the nature of radical politics in the 1780s and 1790s rather well. It would be wrong to assume that radicals accurately represented the political opinions of the middle classes as a whole. But Kramnick is quite right to say that they claimed to do so, and thus right to say that their attack on ascribed rank was "a self-conscious glorification of the mission of the middle class in English society."[34] Radical intellectuals of the 1790s were struggling to define a coherent middle-class political language that did not yet exist.[35] They did this in large part by adapting existing republican traditions. Although those traditions were in principle agrarian, they had already undergone some stretching to encompass the political practice of an oligarchic and mercantile state. Whether associated with Country opposition to Walpole, or with a broader Commonwealth tradition, most forms of republican idealism in the first half of the eighteenth century shared the assumption that the security of a republic depends on the presence of a propertied political class whose members are materially self-sufficient and therefore free to act in the public interest. Not all forms of property seemed equally well adapted to this mission. Land was traditionally conceived as an inalienable personal inheritance, supporting an equally inalienable political personality: "it anchored the individual in the structure of power and virtue, and liberated him to practice these as activities."[36] Mobile property was not understood to support the same kind of autonomy. Instead of liberating its possessor for leisure and civic virtue, it might tempt him to alienate his civic capacities by selling or buying votes, or by paying a standing army to defend the republic.[37]

But political practice in eighteenth-century Britain was dominated by networks of patronage that controlled parliament; under this oligarchic system, power depended at least as much on control over flows of (public and private) money as on ownership of land. The tension between political ideals and political practice generated indictments of corruption, stockjobbing, and place-mongering. But it also promoted strategically vague definitions of "independence" and "property" that could stretch to include substantial merchants, or landowners whose wealth in fact rested on substantial interests in trade.[38] Early eighteenth-century society was polarized along a number of geographical, social, and political boundaries, but it was not sharply divided into an agricultural and a trading or manufacturing class. Though the tradition of republican idealism was ostensibly agrarian, it tended in fact to generate a zone of ideological consensus between the gentry and the upper ranks of the middle class, which could agree at least that property of some kind or another was a precondition for political responsibility.

By the 1760s, this consensus was growing unstable. Traditional concepts of rank, order, and estate were beginning to give way to a new conception of "class" that sorted individuals according to their mode of producing or consuming economic value.[39] Though writers continued to refer to the "middle ranks" and "middle orders," the political claims made in behalf of those groups were increasingly underwritten by a new moral emphasis on productivity, and thus by implicit reliance on the category of class.

Writers like Henry Brooke, for instance, suggested that work, rather than property itself, was the foundation of political responsibility.

He who labours hard to acquire a property, will struggle hard to preserve it, and exercise will make him active, robust, and able for the purpose. As the man of industry hath in himself a living fund of competence for his own occasions, he will be the less tempted to plunder or prey upon others; and the poignant sense and apprehension of being deprived of a property, so justly acquired, will give him the nicer and stronger sense of such an injury to others.[40]

Political responsibility here depends on possession of a "living fund of competence." Though the words *fund* and *competence* evoke property, there is a revealing play of words in the expression. Since labor has made the man of industry "active, robust, and able," he has "*in himself* a living fund" more secure than any estate. The pun offers a miniature model of the process whereby the middle classes appropriated the Country–Commonwealth ideal of independence, and stretched it to cover class fractions (for instance, artisans and professionals) that had not been well served by earlier eighteenth-century idealizations of property. Though political responsibility is still founded on moral independence, independence is now defined as spontaneity rather than self-sufficiency: it appears as the free choice to work, not as leisured freedom from want. In the 1770s, as James Burgh, John Cartwright, and the Society for Constitutional Information argued for a reform of parliament that would extend the franchise to all male citizens who worked and paid taxes, old political conflicts were reorganized around explicitly economic boundaries of class. The redefinition of autonomy suggested by Brooke's pun became increasingly important to opposition politics.[41] Independence was still a requisite for political virtue, but independence was now to be understood, not as civic-spirited leisure guaranteed by property, but as the activity of work itself.[42]

The chief function of *energy* in political novels of the 1790s is to underline this redefinition of autonomy. *Energy* becomes central in these novels because it operates as shorthand for talent and political virtue at once, and because it implicitly identifies both things as the spontaneous produce of the human mind. Thomas Holcroft is perhaps best known now as one of the four defendants in the treason trials of 1794; his novel *Hugh Trevor* (1794–1797) was in fact interrupted for three years by the trials. In his preface to the book Holcroft states that "the interesting question agitated in the following work" has two parts: first, "the choice of a profession," and secondly, the "growth of intellect"—more specifically, the shaping of natural gifts by experience.[43] These social and psychological themes are in practice fused, because *Hugh Trevor* is structured around a conflict between the protagonist's desire to display his own natural gifts, and his slowly growing experience of the corrupt state of existing society. Entrenched aristocratic interests thwart Hugh's ambition successively in the church, in political writing, and in the profession of law.

Energy enters the novel as a one-word summary of Hugh's natural moral and intellectual potential. Mr. Turl, who acts as Hugh's preceptor and model, comments on reading Hugh's first manuscript that his style, though over-embellished and otherwise flawed, is possessed of "the first great quality of genius . . . Energy."[44] Hugh emphasizes his trust in the same quality as he sets off for London to choose a profession for the first time, and enter adult life.

> Under all points of view, my constant hope was in the energy of my own mind. Among the numerous examples which I had seen, of men who had gained preferment, many by the sole influence of personal interest, and many more by the industry of intriguing vice, there were some who had attained that end by the exertion of extraordinary talents and virtue. It is true they were but few, very few; yet on them my attention was constantly fixed. Them I was determined to emulate, exert the same powers, rise by the same means, and enjoy the same privileges.[45]

The opposition set up here between different means of achieving "preferment"—rising through interest and rising through merit—is central to the novel. But the notion of merit involved needs to be specified more closely. Merit is not described here in terms of the traditional middle-class virtues of industry and thrift. The contrast between "the exertion of extraordinary talents and virtue" and the "industry of intriguing vice" seems to suggest something slightly suspicious about "industry" alone, as if effort could only be meritorious when it exercises an inherent, natural endowment of talent. Thus the advantage of the rubric that frames the passage: "the energy of my own mind." *Energy* can combine talent and exertion in a single word because it is a quality of mind that, by definition, includes its own spontaneous exercise.

William Godwin's early political novel *Caleb Williams* also relies on *energy* to represent the natural talents that go wasted and unrewarded under an aristocratic system. The clearest instance of this appears in Caleb Williams's discussion of the bandits led by Mr. Raymond.

> Uninvolved in the debilitating routine of human affairs, they frequently displayed an energy, which from every impartial observer would have extorted veneration. Energy is perhaps of all qualities the most valuable; and a just political system would possess the means of extracting from it thus

circumstanced its beneficial qualities, instead of consigning it as now to indiscriminate destruction. We act like the chymist who should reject the finest ore, and employ none but what was sufficiently debased to fit it immediately for the vilest uses.[46]

The failure of aristocratic society to extract human energy is represented here as a perverse failure to appreciate a natural resource.

The meritocratic arguments of Godwin and Holcroft suggest that *energy* had a special utility for middle-class intellectuals who believed their fortunes were disproportioned to their talents. It is tempting to connect this fact to Paul Keen's description of an ambivalence that already separated "the professional and commercial sectors of the middle class." Though both groups believed that commerce could polish manners and foster liberty, professionals tended to temper their enthusiasm for commerce with remnants of a republican and/or Protestant distrust of luxury. Professional writers— and Keen hints, professionals in general—looked to the "republic of letters" for a form of individual ambition that seemed easier to reconcile with republican ideals of public virtue.[47] Was energy, then, a distinctively professional ideal, perhaps related to modern notions of professional autonomy?

Keen is entirely right to suggest that different fractions of the middle classes assigned different weight to economic and symbolic capital. But an idealization of work as spontaneous activity was one of the beliefs that united these different groups, and permitted them to declare common cause under the umbrella of "the middle ranks of life." It is, by contrast, very hard to locate a coherent professional identity in late-eighteenth-century Britain. Anglican ministers and barristers felt little common cause with surgeons, or attorneys, or with so-called professional writers. The upper and lower branches of the professions occupied distinct social worlds. Barristers, physicians, and the clergy were linked to the gentry—in part by social ties to their clients, and in part through the requirement of a classical education. What status they possessed came from a claim to gentility. It did not come from an ideal of professionalism, because there was no system of professional education or certification, and thus no foundation for a specifically professional form of symbolic capital that could be distinguished from dress, elocution, social connection, and command of Latin.[48] The status of attorneys, apothecaries, and surgeons—who could not claim gentility—was accordingly much lower than that of barristers and physicians. The impetus toward professionalization in the modern sense (the creation of legally recognized monopolies on training and on competence) came, naturally enough, from these lower branches. As W. J. Reader points out, the apothecaries were the first to open the path: the passage of the Apothecaries' Act in 1815 "marked the emergence of the nineteenth-century general practitioner, in the sense of a practitioner holding recognized qualifications in medicine and surgery."[49]

I dwell on this point, at the risk of belaboring it a little, because professionalism is becoming a dominant rubric in the study of Romantic literature and history. The word *professional* appears in a growing number of titles, and Clifford Siskin has undertaken the impressive task of reimagining Romantic literary history in terms of the emergence of "disciplinarity" and professionalism. Siskin freely acknowledges that he is describing the emergence of social categories that were not yet dominant, and it is important to keep that fact in mind.[50] Since the distinctive autonomy associated with professional occupations depends on those professions' power to control the accreditation of their members, it is largely anachronistic to talk about professional autonomy in the 1790s, when no such power existed. *Energy* expressed an ideal of autonomy, but one linked instead to the broader social category of class. Its appeal extended from an engraver like William Blake to a teacher like Mary Wollstonecraft, and was founded on a promise to reconcile freedom with productive labor. Though middle-class artisans, tradesmen, and professionals in fact remained dependent on a network of clients and patrons, invocations of *energy* could represent their work as a politically relevant form of independence.

Energy Versus "Mechanical and Daily Labor"

By 1798, the word *energy* had become a recognizable tic in British radical prose; it was accordingly taken up by young Tories as a central element of their parodic attack on radical ideas. In these parodies, *energy* typically links the Godwinian tenet of "perfectability" to overweening middle-class ambition. In their preface to "The Loves of the Triangles," George Canning and John Frere make their imaginary author, Mr. Higgins, gloss perfectibility as the belief that "we have risen from a level with the *cabbages of the field* to our present comparatively intelligent and dignified state of existence, by the mere exertion of our own *energies*. . . ." Godwin's belief in historical progress is combined with Erasmus Darwin's belief in evolution, and both are made to look like the self-congratulatory puffing of a middle-class social climber. In the end, Mr. Higgins thinks, man will be so independent that he need "never die, but *by his own consent*."[51]

The sincere, but confused, British radicals in Elizabeth Hamilton's novel *Memoirs of Modern Philosophers* are similarly characterized by their desire to rise through personal energy. Mr. Glib, in particular, is given the set speech "That's it! Energies do all!" as his leitmotif. To a man whose finger has just been bitten by a monkey, Mr. Glib cries, "Exert your energies, my dear citizen . . . exert your energies, my dear. That's it! Energies do all! Cure your finger in a twinkling. Energies would make a man of the monkey himself in a fortnight."[52] Mr. Glib's pupil Bridgetina has absorbed similar ideas from her teacher. She looks forward to the era "when mankind are sufficiently enlightened to cure all diseases by the exertion of their energies."[53] Godwin's utopian political ideas are thus represented as resting on a foundation of quack medicine.[54]

But Hamilton has another satirical strategy. Her radical philosophers refer to energies not just to underwrite their quackish idea of perfectibility, but also to justify their evasion of several traditional moral duties—in particular, the duty to work. This satirical point was present, in fact, from the very inception of Hamilton's attack on Godwinian energies. Her *Translation of the Letters of a Hindoo Rajah* (1796) already contained a brief caricature of Godwin as a "Mr. Vapour," an atheist who worships "circumstances, energies, and powers" as his gods. According to Mr. Vapour, when the millennium—which he calls "The Age of Reason"—arrives,

> The fear of punishment . . . that ignoble bondage, which at present restrains the energies of so many great men, will no longer damp the noble ardour of the daring robber, or the midnight thief. Nor will any man then be degraded by working for another. The divine energies of the soul will not then be stifled by labouring for support.[55]

The opposition between "energies" and "labouring for support" established in this last sentence is a crucial one. For all of Hamilton's philosophers, energies differ from mere labor in being utterly spontaneous and, indeed, uncontrollable. The same point is made in *Memoirs of Modern Philosophers*, when Mr. Glib exclaims, "Nothing so bad for energies as order: eat when I please, sleep when I have a mind. That's it! my dear, that's the way to have energies."[56] The spontaneity of energy expresses an ideal of independence taken to its logical (that is to say, anarchic) extreme. On discovering a treatise about the "Gonoquais Hottentots," Mr. Glib finds their supposedly anarchic and pre-governmental condition a perfect image of his own desire for energy. "See here, Citizen Myope, all our wishes fulfilled! All our theory realized! Here is a whole nation of philosophers, all as wise as ourselves! . . . No government! No coercion! Every one exerting his energies as he pleases!"[57] This critique of Godwin's utopia as a state of primitive anarchy is underlined by the direct echo of Godwin in Myope's response.

> "Yes," said Mr. Myope, "and as we well know mechanical and daily labor to be the deadliest foe to all that is great and admirable in the human mind, to what a glorious height of metaphysical knowledge may we expect a people to soar, where all are equally poor and equally idle."[58]

The irony at Mr. Myope's expense is heavy-handed, but the first part of the sentence does precisely echo Godwin's statement, in *The Enquirer* (1797), that "mechanical and daily labor is the deadliest foe to all that is great and admirable in the human mind" (*GW* 5:154).

In fact, Hamilton was by no means attacking a straw target. A firm opposition between the free deployment of human energy and "mechanical and daily labour" undergirds much of Godwin's writing. The book "Of Property" in *Political Justice* suggests that in an ideal society, manual work would be reduced to half-an-hour a day so that "every man would have ample leisure for the noble energies of mind" (*GW* 3:440). The same argument recurred later in two essays in *The Enquirer*—"Of Riches and Poverty" and "Of Avarice and Profusion." Godwin's aim in both essays is to resist the received idea that labor, or "industry," is a moral good in itself. Although "industry has been thought a pleasing spectacle," Godwin argues that "the genuine wealth of man is leisure," because "those hours which are not required for the production of the necessaries of life, may be devoted to the cultivation of the understanding . . . " (*GW* 5:155, 5:153, 5:156).

Godwin is always careful not to exalt idleness as such over activity. The distinction between industry and energy is rather a distinction between two different forms of productive activity. In part, it is based on the old valorization of mental work over bodily work, since "that which we possess in common with the brutes, is not of so great value, as that which we possess distinctively to ourselves" (*GW* 5:184). But the distinction is more fundamentally a consequence of Godwin's insistence that human beings should have absolute self-determination. In his essay "Of Self-Denial," Godwin states that "true energy" depends on independence. Independence in turn depends on a willingness to do without sensual luxuries; but Godwin goes on to say that it is a mistake to think of independence and energy as purely intellectual states. Sensual pleasures and bodily exertion also play a role in the ideal. "He that would have great energy, cannot do better than to busy himself in various directions, and to cultivate every part of his nature." The energetic man should have "a sound body, as well as a sound mind," and should even deliberately cultivate the senses, the "digestic powers," and the "animal œconomy annexed to the commerce of the sexes" (*GW* 5:183, 5:185).

The Tension Between *Energy* and *Industry*

Godwin was by no means the only writer to distinguish self-determining energy from mechanical and daily labor. He is only one instance of a widespread tendency in the 1790s to distinguish between two kinds of production: production that wells up spontaneously in the worker, and production that is merely forced or willed. Old terms of praise that suggested meticulous activity were coming into disrepute. Instead of praising diligence or industry, writers preferred to praise a productive energy that flowed spontaneously from an inner source. Contempt for the laborious elements of society was nothing new; it was a long-standing element of aristocratic ideology, since (as Stephen Greenblatt among others has shown) the very identity of the aristocrat depends on a displacement of toil onto others.[59] In the late eighteenth century, however, this contempt for painstaking labor was incorporated into middle-class ideology—and became a central tenet of bourgeois radicalism—where it accompanied, paradoxically enough, an assertive work ethic. This was possible because of a newly systematic distinction between the spontaneous energies of free (i.e., middle-class) workers and the mechanic toil of laboring drudges.

When Godwin made this sort of distinction, it had a leveling force: the point was to show that mechanical labor was incompatible with human dignity, and that the material prerequisites of middle-class independence should therefore be extended to all social orders. But other writers deployed the same distinction in order to gentrify the work ethic, and distinguish it from plebeian toil. This was especially true of writers who, like Richard Payne Knight (1750–1824), straddled the boundary between middle-class and aristocratic worlds. Knight's grandfather had been an ironmaster in Shropshire; Knight himself was a collector of antiquities and aesthetic theorist, who in 1772 had Downton Castle built in a picturesque mixture of Gothic and Grecian styles.[60] Knight's poem *The Progress of Civil Society* (1796) shows how deeply the opposition between spontaneous work and deliberate toil could be felt. In this long philosophic poem, Knight uses *energy* frequently and in a variety of contexts: most notoriously, to complain that the indissolubility of marriage numbs and cramps "the soul's best energies"—an association of energy and libertinage that prompted a parody in the *Anti-Jacobin*. But what interests me at the moment is Knight's use of *energy* to describe the spontaneity of labor. This begins in the introduction, where Knight hopes to imitate Lucretius's verse, which "flows equal to the occasion, as the spontaneous efflux of a mind of such vast and universal energy, as never to need any extraordinary exertion." This statement already contrasts two styles of work. One flows, "vast and universal," "clear, simple, elegant, and vigorous," and the other requires "extraordinary exertion." The implication is that the highest achievement takes place without exertion, as a spontaneous product of the mind's natural energy.[61]

This idea develops more complexity in the poem's fourth book, "Of Arts, Manufactures, and Commerce." Knight has already traced human society through the stages of "hunting," "pasturage," and "agriculture." His fourth book seeks to explain the origin of money, of commerce, and of fine gradations of social class. The justification of class inequality he offers is a common one in the period.[62] Inequalities in society serve the same function as disequilibrium in a machine: that is to say, they keep it moving. "So in well-poized and complicated states, / The separate classes act as springs and weights." What needs to be kept moving is, preeminently, money, "the unequal currents of the golden tide / [which] still, as more dispersed it ebbs and flows, / enlarged in energy and substance grows."[63]

But matters become more complex, for wealth has its perils, chief among them satiation. The "golden tide" that is possessed of such energy paradoxically betrays its possessors to dissatisfaction and "listless languor of . . . soul." Knight describes various ways the wealthy try to defeat this dissatisfaction—including "savage sports" and affectations of pastoral simplicity—but concludes that without intellectual activity these distractions are doomed to be hollow. Try as they will, the rich will still be "without the vigorous nerves, that brace / The savage hunter for the mountain chase; / And void of all that energy of soul, / Which springs unchecked by mortal man's control." The energy of money seems inevitably to evoke an opposite and countervailing principle, that of languor. By destroying spontaneity, the energy of commerce destroys itself.[64]

This paradox casts light on a difficulty the ethic of mental energy presented to those writers who adopted it: how to explain the origin of vice? At first it might seem that the answer would be simple. If energy is good, then blameworthy things must originate in a want of energy. That explanation will not do, however, because man has already been defined as essentially energetic, and moreover because the political point of energy lies in its spontaneity. To blame a want of energy might quickly translate into blaming indolence—and yet writers like Knight are precisely not calling for an ethic of self-regulation. For a similar reason Humphry Davy hesitated in a notebook I have quoted above, writing that man is an "an active energetic being" whose "miseries in a great measure depend upon ["his indol" crossed out] himself."[65] An essentially energetic being can hardly be blamed for indolence. The solution adopted by Davy and Knight is to find in energy itself the seeds of its own corruption. As Davy wrote in an unpublished essay on genius: "Genius, dangerous genius accidentally exposed to new impressions had directed one part of her energies towards procuring new modes of exciting the languid senses of civilized man. Hence has arisen Luxury—hence commerce—hence unnecessary labor. . . ."[66] Spontaneous energy, through its superabundant production of luxuries, leads to the unnecessary labor that is its own polar opposite. Davy involves himself in a paradox here by making "the languid senses of civilized man" precede the very "Luxury" that is supposed to call them into being. Some such paradox is always involved, however, in an attempt to unfold a morality of vice and virtue out of a state of original innocence. The specific state of innocence claimed by the language of energy is a state before Adam's curse. One can read these texts by Davy and Knight as different versions of a proof that, in the beginning, work was not bound up with pain, and attempts to explain the fall by which energetic creation degenerated into painstaking labor. In Knight's case (and also, incidentally, in Davy's) this historical myth is related to an account of the difference between the West and the Orient. Knight moves onto this terrain by explaining that the debility of the rich is just one instance of a general point about human nature, which is that

> In social trammels too severely train'd,
> Man finds each brighter faculty restrain'd,
> And every genuine spark of native fire,
> Dimm'd and compress'd in frigid rules, expire.[67]

By writing "social trammels" Knight includes all sorts of constraints, but what he has chiefly in mind are impermeable class boundaries. This becomes evident when he goes on to contrast "the swarms, which China's fertile soil / Maintains in slothful ease, or abject toil" against an ideal realm, "to art and genius kind, / Where social ranks are mark'd, but not confin'd." Knight goes on to describe this ideal society as a place where property rights are secured, but avenues for advancement are left open "to stimulate the forces of the mind." He is of course idealizing those aspects of British society that already came close to constituting a liberal state.[68]

This state finds its model in the "wild tumults and intestine jars" of ancient Greece. Here conflict and the spirit of emulation led to an energy that somehow managed to remain genuine instead of sliding into languor. Greek achievements are real—that is to say spontaneous—achievements.

> By no dull methods cramp'd, or rules confined,
> Each effort bore the impression of the mind;—
> Warm from the fancy, that conceived it, flow'd,
> And, stamp'd with nature's genuine image, glow'd.

To be spontaneous is also, of course, to be natural. Knight opposes these qualities to an Orient onto which he projects the qualities of artificiality and luxury.

> With scorn behold the unactive Mede enfold
> His languid limbs in silks enwrought in gold;
> And Egypt's sons, in frigid method bound,
> Still onward move their dull mechanic round.

The opposition between energy and languor is, oddly enough, identical to the opposition unfolded in Knight's introduction between "vast and universal energy" and "extraordinary exertion." To work in a "dull mechanic" way is, for Knight's purposes, equivalent to being "unactive." Knight calls exertion "peaceful drudgery," or "patient, dull, mechanic labor," and represents it as a form of passivity.[69]

Rhetoric like this tends to qualify Paul Keen's observation that the middle classes sought in the 1790s to legitimate themselves through an "industrious self-image." Keen persuasively reconciles republican and bourgeois-radical interpretations of the period by remarking that, for bourgeois radicals, work came to fill the place of republican virtue. "The political emphasis on the moral independence of the individual remained intact; it was simply being redefined in terms of the individual's integration within, rather than distance from, the relations of production."[70] But to this, one should add that the representation of work as independence involved a concomitant division between different kinds of work. Few writers could believe that domestic service or hired agricultural labor provided a foundation for moral independence. Those occupations were too obviously forms of servitude. Even radicals who endorsed universal manhood suffrage expressed some reservation about its extension to "menial servants," for instance, because the labor of servants seemed to entail "dependence on their masters."[71] The function of a term like *energy* was to distinguish the kind of work that ought to count as independence from the mechanic toil of the laboring classes. It did so, not by acknowledging the social relations that actually divided different kinds of work (ownership of the physical or symbolic means of production), but by reifying independence as a spontaneity analogous to the spontaneity of natural forces. The *energy* of spontaneous work was thus opposed to "peaceful drudgery" or "mechanical and daily labor," and its relationship to older terms of praise like *diligence* and *industry* became at best ambiguous.[72]

The troubled relation between *energy* and *industry* was rooted in the conflicting impulses of middle-class writers who sought simultaneously to claim work as a proof of social responsibility, and to disavow it as a constraint on individual freedom. Romantic equivocations about the spontaneity of poetic production use slightly different language, but worry the same underlying problem. With the "spontaneous overflow of powerful emotion recollected in tranquility," Wordsworth compresses the equivocation into a single phrase. Keats unravels it more gradually by postulating that poetry should come as naturally as the "leaves to a tree"—and then retracting that assertion piecemeal in subsequent letters and poems. Chapters 3 through 6 will examine the ideal of spontaneous literary production in several Romantic-era poets, asking how that ideal was related to writers' sense of class identity.

CHAPTER 3

APOLLO, GOD OF ENTERPRISE

The Romantic revival of polytheism was selective. Zeus, Hera, Ares, and even the nine Muses remained antique names; the sun-gods, on the other hand (Hyperion, Apollo, Helios, and, as a kind of mediator with the sun, Prometheus) became favorite protagonists. The disproportionate emphasis on the sun extended beyond Hellenism: British Romantic writers also developed a strain of Zoroastrian poetry that interpreted Ormusd (or Oromazde) as a god of fire and sunlight.[1] The names of Peruvian and Egyptian sun-gods could be added to the list, if listing deities weren't finally beside the point. Romantic mythographers, including Charles-François Dupuis and William Drummond, transformed all gods into sun-gods, by arguing that religion itself was originally sun-worship. The religious impulse, in Dupuis' account, springs from man's gratitude for light's "creative energy"—"the true principle of our existence, without which our life would only be a sensation of protracted weariness."[2] Though the original meaning may have been covered up or forgotten, all religious stories— including the birth and death of Christ—were according to Dupuis once representations of the sun's daily journey across the sky or yearly journey through the stations of the zodiac.

As Marilyn Butler has pointed out, critics of Christianity found this argument useful. In the works of Byron and P. B. Shelley, solar myth often advances a skeptical agenda, implicitly demystifying religion by translating it back into its original physical form.[3] But the solar critique of mythology had a tendency to become itself mythopoeic; in Shelley's "Hymn of Apollo," Keats's "God of the Golden Bow," and Manfred's address to the sun, one finds avowals of dependence as earnestly rhapsodic as anything in Christian poetry. Nor is it easy to separate naturalistic invocations of solar myth from other writers' efforts to develop Christianity itself into a solar religion. Christ and Father Helios rub shoulders in Friedrich Hölderlin's "Patmos."[4] For William Blake, the Incarnation and Passion are portions of the fall and redemption of Los (who is, among other things, an anagram of Sol and the creator of the sun). S. T. Coleridge's "Religious Musings" (1794–1796) compare the birth of Jesus to the creation of the solar orb; the redemptive power of Christ becomes the sun's ability to melt a stream in winter, and the Love through which the poet lives and writes becomes the sun's all-enlivening warmth.[5] These metaphors fit comfortably into Christian iconography; Milton's invocation of "holy light" in Book III of *Paradise Lost* ("since God is light / And never but in unapproached light / Dwelt from Eternitie") might be cited as a proximate source.[6] But when half of the Romantic sun-gods are explained as examples of Christian tradition, and the other half as examples of anti-Christian polemic, one begins to suspect that neither hypothesis is doing much to explain the phenomenon. Romantic writers' penchant for a solar interpretation of religion would appear to have some broader foundation.

One thing Christian and neo-pagan texts had in common was a tendency to invoke the sun as a synecdoche for the vitality and agency of the physical world as a whole. A song by Thomas Moore, praising the Christian god in terms of an extended analogy to the sun, begins "Thou art, O God, the life and light / Of all this wondrous world we see. . . ."[7] In P. B. Shelley's "Hymn of Apollo," the Greek sun-god claims something similar: "Whatever lamps on Earth or Heaven may shine / Are portions of one power, which is mine" (*SP* 613). One might explain the prominence of sun-gods in Romantic poetry, then, as an aspect of poets' dynamic naturalism—their habit of invoking nature in the form of a ubiquitous force, which emblematizes nature's universality because it flows through all action.

Eighteenth-century natural philosophy was largely responsible for this view of nature. As I mentioned in chapter 1, Isaac Newton called for causal principles that were framed "not as occult Qualities, supposed to result from the specifick Forms of things, but as general Laws of Nature, by which the Things themselves are form'd."[8] That instruction tended to reduce the prestige of taxonomic explanation (nature as a field of differentiations) and increase the prestige of dynamic modeling (nature as an interplay of forces). Following Newton's example in the *Opticks*, eighteenth-century natural philosophers came to expect that all the qualities and forms of nature—even apparently "primary" qualities like figure and solidity—could be explained by forces that were in themselves formless.[9] Light, heat, and electricity were the leading candidates, and natural philosophers believed that all three of those forces could be traced back to the sun. The sun therefore became a visible emblem of the invisible and ubiquitous force(s) contained in all life and motion. It was no longer understood to organize the terrestrial world from above. But it acquired a position of even greater importance as the one "specifick Form" that could act as a synecdoche for "general Laws"—or as G. W. F. Hegel put it, the individual form privileged to signify nature's universality. "[T]he most universal physical element" is light, "but though possessing thus in itself universality, it exists at the same time as an individuality in the Sun."[10]

Tracing Romantic solar myth back to its sources in natural philosophy does not by itself explain the period's fascination with sun-gods. It defers the question. If solar myth turns out to be a figure for what I have called "dynamic naturalism," one next has to ask why Romantic writers—Christians and skeptics alike—were so profoundly obsessed with dynamic naturalism. In chapter 2, I began to offer an answer to that question, by proposing that late-eighteenth-century British idealizations of natural force had as much to do with politics as with philosophy. "Energy" became a charged code word in the 1790s because middle-class radicals were fond of the analogy it implied between the autonomous work of the "middle ranks of society" and the spontaneously active forces of nature. By identifying autonomous work as the only valid basis for social distinction, middle-class radicals both distinguished themselves from the laboring drudges beneath them, and undermined the theory of agrarian independence that had justified the landed gentry's monopoly on political power. Real independence came not from ownership of land but from a spirit of self-directed activity analogous to nature's spontaneous energies. This chapter argues that analogies between dynamic naturalism and middle-class autonomy also shaped Romantic interest in solar myth. A few writers invoked the sun as a synecdoche for natural force because they were interested in natural force itself, but many were more directly interested in the political construction that had been put on the concept, and some (like William Blake) borrowed the political symbolism that had accreted around solar force without being particularly enamored of natural philosophy or even of the abstraction "nature."

Taking these qualifications (and a few others) into account, a more adequate explanation of British Romantic heliolatry might look something like this: Invocations of the sun in British Romantic writing tend to be ways of reconciling idealization of autonomy with a minimal acknowledgment that individuals belong to a world larger than themselves. In republican tradition, independence had been an explicitly civic ideal. Although middle-class revisions of that ideal tended to minimize its foundation in property and idealize absolute spontaneity, they retained a collective resonance—at least the kind of displaced collectivity that is implied by any effort to situate personal independence within a larger natural order. There was something paradoxical, however, about tracing spontaneity to a celestial body or a god, or even to a diffuse abstraction like nature. Any higher power that ratifies personal independence simultaneously seems to undermine it.

Romantic writers were aware of this paradox; the famously circular theory of the imagination found in Wordsworth's *Prelude* or Shelley's "Mont Blanc"—in which nature teaches the mind how to be independent of nature—solves the dilemma by embracing the contradiction. Solar imagery provided another way to embrace it. Because eighteenth-century natural philosophy had changed the sun from a specific organizing form into an emblem of universal and placeless agency, to apostrophize the sun was in effect to apostrophize the spirit of autonomous agency itself. By invoking solar science and solar myth, writers could represent liberty and energy as aspects of nature (and thus as public and collective ideals), without compromising their claim to autonomy.

I have characterized this as a middle-class argument, and it is true that the struggle to displace property and ascribed rank with a strictly dynamic conception of merit was advanced, in the 1790s, by middle-class radicals. When Mary Wollstonecraft asks, "how can a rational creature be ennobled by anything that is not obtained by its *own* exertions?" she expects her audience to understand the implications for both gender and class: "Abilities and virtues are absolutely necessary to raise men from the middle rank of life into notice, and the natural consequence is notorious—the middle rank contains most virtue and abilities."[11] But when writers translated this middle-class argument into solar terms, the point of the translation was often to universalize the argument, and disavow its class specificity. Wollstonecraft states forthrightly that all women should have the opportunity presently granted to middle-class men—"an opportunity of exerting themselves with dignity, and of rising by the exertions which really improve a rational creature."[12] But Mary Robinson prefers to evade class specificity (as I will shortly show) by representing middle-class ambition as a spontaneous natural force—an "effulgent ray," rather than a form of exertion. This universalizes middle-class ideals, but also gentrifies them. Although Wollstonecraft was the better writer, Robinson's rhetoric was more successful; it universalized itself so effectively that—as chapter 6 shows—it was eventually assimilated by writers from the gentry and aristocracy. At that point one should begin to call it a liberal rather than a specifically middle-class invocation of natural force.

The studies of individual writers in this chapter have been organized in a sequence that tests what I expect will be the most controversial part of the thesis outlined above: my claim that mythopoeic descriptions of Prometheus and Los and Apollo need to be linked to, and understood in the context of, other works that made a more explicitly physical connection between sunlight and economic autonomy. At present, critics tend to assume that the liberal argument of Romantic heliolatry was latent in ancient myth and only needed the right circumstances to emerge: that seems to be implied, at any rate, when critics use adjectives like *Promethean* or *Apollonian* to describe an ethic of self-improvement and social aspiration. Though the myth of Prometheus affirmed human beings as restless and forward-looking creatures, it was not at first a story about liberal individualism—and especially not a story about the analogy between individual aspiration and the pure, commutable agency of sunlight. When critics use *Promethean* or *Apollonian* in the latter sense, they are talking about an idea constructed in the late eighteenth century.

The history of that construction tends to be obscured by the fact that the most famous examples of Romantic heliolatry come from a relatively late stage in its development. By the Regency period, writers like John Keats and P. B. Shelley had developed a virtuosic ability to transpose modern arguments into an imaginatively consistent ancient world. But Keats and Shelley were building on a tradition of political symbolism that had developed in the 1790s, in texts where the names of Prometheus, Apollo, and Mithras were grace notes ornamenting a naturalistic and frankly contemporary analogy between sunlight and individual liberty. To restore that neglected context, this chapter will move from poems about the physical sun to poems about solar gods, and build a bridge between the two. I begin by contrasting James Thomson's "Liberty" (1735–1736) to Mary Robinson's "Progress of Liberty" (1798). Both poems invoke the sun as a symbol of British liberty, but they describe liberty very differently, and the difference reveals how closely political discourse in this period is bound up with changing representations of nature. The next section, on a poem by Humphry Davy, shows how one Romantic scientist represented sunlight simultaneously as a physical power, as subjective experience in general, and as middle-class ambition in particular. The third section begins to build a bridge from these political invocations of the sun to solar mythography, by considering Apollo's role as a god of middle-class liberty in Henry Boyd's historical drama *The Helots* (1793). The final two sections of the chapter consider how the fusion of dynamic naturalism, middle-class consciousness, and solar myth in Boyd may cast light on the better-known solar deities of Blake and Keats.

The Sun as an Emblem of Middle-Class Energy

British poets used the sun as a figure for liberty throughout the eighteenth century. But the figure grew more prominent, and developed a specific appeal for middle-class writers, toward the century's end. That change reflects a broader transformation of political discourse in eighteenth-century Britain, which was (from about 1770 forward) increasingly organized around class—a relatively new social category that divided society not according to location, or ascribed rank, but according to the functions performed by different groups in the production of economic value. One consequence of this transformation was to shift attention from the relationship between political liberty and propertied leisure, to an analogy between liberty and production—an analogy that was particularly valuable for the tradesmen, artisans, and professionals who were becoming conscious of themselves as an industrious, taxpaying, but disenfranchised class of persons. When early-eighteenth-century poets compared liberty to the sun, they had been primarily interested in the sun's ability to inform and organize a system of freely moving bodies. That metaphor allowed them to explore the paradox that political freedom both depends on and produces an orderly distribution of property and status. At the end of the century, the analogy focuses on sunlight itself, conceived as a force that directly animates and enlightens human beings. The thrust of the analogy between liberty and sunlight is now to imply that all forms of human activity—commerce, social feeling, genius, and love—are instances of liberty, just as all forms of human energy are instances of sunlight. This undermines the agrarian argument that possession of land is a precondition for political responsibility, at the cost of blurring boundaries between the political and economic realms, or for that matter between politics and physiology.

In early eighteenth-century Britain the discourse of political opposition relied on an agrarian logic, because real property was seen as the most reliable foundation for public virtue. But the agrarian rhetoric of republican tradition cannot always be taken to imply a defense of specifically landed and aristocratic interests. The idealization of land as political independence obviously served the interests of the aristocracy and gentry, but it also encompassed small freeholders (in principle), and in the first part of the eighteenth century it was often in practice endorsed by members of the urban middle classes. In the period of Walpole's ascendancy, for instance, idealization of the independent landowner became a central theme of opposition to the court's monopoly on patronage and control of Parliament. It was thus a theme that united Tory landowners and urban Old Whigs who felt disenfranchised by the existing oligarchy.[13] Though the paradigmatic independent citizen might be a landowner, Opposition writers often left terms like "independence" and "property" vague enough to apply, at least by analogy, to substantial merchants as well.

Equivocation about the material basis of independence is visible, for instance, in James Thomson's long progress poem "Liberty" (1735–1736), which follows the goddess of Liberty from her origins in pastoral society to her apotheosis in Great Britain. The poem was dedicated to Prince Frederick at a time when he was flirting with (and about to join) opposition to Walpole. Thomson faithfully reproduces Opposition ideology in founding liberty on the virtues supposed to characterize independent landowners: "independent life; / Integrity in office; and, o'er all / Supreme, a passion for the commonweal."[14] But Thomson (himself an urban Whig) leaves his depiction of "independent life" strategically vague; the "private field" he describes, whose "happy master" is "the only freeman," wavers between a specifically agricultural field and a figurative representation of property in general.[15] As Pocock observes, Country–Commonwealth ideology rarely vilified trade itself.[16] Since the boundary between the landowning and trading classes was by no means clear in Britain, uneasiness about the political role of mobile property tended to be displaced instead onto placeholders and speculators in public credit. Qualities of thrift and moderation, on the other hand, could transform a fluctuating income into a personal endowment, and make mobile property an adequately secure foundation for civic independence. Thomson accordingly spends more time celebrating "true-judging moderate desires, / Economy and taste" than he does talking about plowshares that become swords.[17] The early-eighteenth-century idealization of propertied independence was in practice an unequal compromise between the middle classes and the landed gentry. Since the rhetoric was agrarian, the gentry got the better half of this ideological bargain. But the compromise expressed a shared sense that liberty depended on permanent freedom from want, which could only be founded on possession of property that was, if not always realized as land, at least conceived as a stable and inheritable personal endowment.

The solar imagery of Thomson's "Liberty" echoes the poem's vision of liberty as propertied independence. The sun becomes the poem's primary figure for liberty because Thomson understands the sun as an indirect organizing power. Liberty cannot, after all, order her worshippers to take a specific course of action: to do so would contradict her own identity. And yet, an allegorical progress poem needs to represent its central figure as somehow shaping the course of history. Thomson's solution is to represent liberty as an informing light. So, for instance, he apostrophizes the goddess as the "forming light of life! O better sun! / Sun of mankind!" Liberty is "the power, whose vital radiance calls / From the brute mass of man an ordered world."[18] These figures draw on the iatrochemical conception of sunlight discussed in chapter 1. The iatrochemical sun is an organizing and individuating principle; it differentiates gross matter into forms that are each endowed with a shaping virtue or "Governing Spirit."[19] More recent Newtonian theories of gravitation and refraction do make an appearance in Thomson's poem, but those theories are similarly interpreted as ways for the sun of liberty to organize the world by distributing autonomy—for instance, by binding different virtues in a ring like "circling planets."[20]

Thomson's description of liberty as an informing light is entirely appropriate for a politics that founds liberty on material self-sufficiency.

Thomson supposes liberty to depend on a distribution of property that guarantees each member of the political class control over his own livelihood. The sun creates a similar system of ordered autonomies by endowing each living creature with its own vital principle and virtue. There is a suggestion—as one might expect in a poem dedicated to the Prince of Wales—that the order created by the sun of liberty involves systematic gradations of rank. "Britannia's bounded kings" are imagined to act in concert with the sun of liberty and pursue a similar organizing project. By distributing honors, they "raise hid merit, set the alluring light / Of virtue high to view" and thereby "spread that better sun / Which lights up British soul."[21] Like sunlight, limited monarchy organizes the commonwealth to foster independence without constraining the actions of any individual.

Mary Robinson's "Progress of Liberty" (1798) also represents liberty as sunlight. But her solar analogy has implications very different from Thomson's, first of all because she takes it more literally. Though her poem is like Thomson's an allegory, Robinson folds allegory back on itself to produce a characteristically Romantic effect of literalness. Liberty is initially apostrophized as a female figure born in the wilderness, who takes the sun as a foster parent: "Thy sparkling eyes / Snatched radiance from the sun! while ev'ry limb / By custom unrestrain'd, grew firm and strong" (RP 3:2). The sun here is the ordinary physical sun, and liberty's metonymic association with it merely associates her with nature in general. But a few lines further on, the sun-worshipper herself becomes a sun, and the metonym evolves into a metaphor.

> The wild was thy domain! At morn's approach
> Thy bounding form uprose to meet the sun,
> Thyself its proud epitome! For thou,
> Like the vast orb, wert destined to illume
> The mist-encircled world; to warm the soul,
> To call the pow'rs of teeming reason forth,
> And ratify the laws by nature made! (RP 3:3)

As often in Romanticism, metaphor is reinforced by metonymy in order to suggest (in W. K. Wimsatt's words) that "tenor and vehicle . . . are wrought in a parallel process out of the same material."[22] Sunlight is a good symbol of liberty, in other words, not just because liberty psychologically illuminates and warms (metaphor), but because Robinson understands liberty quite literally as a kind of physical energy (metonymy—synecdoche in particular).

Robinson's solar figure also differs from Thomson's because its underlying assumptions are drawn from the physiology of sensibility rather than iatrochemistry. Instead of organizing the world by distributing discrete vital principles, Robinson's sun of liberty directly warms, illuminates, and animates human beings.

> Hail, Liberty sublime! hail godlike pow'r
> Coëval with the skies, to earth new born,
> Thou parent of delight, thou source refin'd
> Of human energy! Thou fountain vast
> From whose immortal stream the soul of man
> Imbibes celestial fervour! (*RP* 3:1)

The celestial "fountain" from which the soul "imbibes" fervor is solar. It may or may not specifically echo the chemical and physiological speculations pursued by Joseph Priestley, Antoine Lavoisier, and Jan Ingenhousz in the 1780s, which had made sunlight the source of terrestrial life and motion. But the metaphor clearly does draw on the widely shared premise that there is some particular nervous fluid or power (variously called "sensibility" or "excitability") that allows human bodies to feel sensations and initiate action. Robinson's sun of liberty no longer organizes society or distributes discrete vital principles; it is instead an animating power that infuses agents with the warm passions and enlightened reason that distinguish true action from mechanical repetition.

As Jerome McGann has pointed out in another context, the concept of sensibility allows Robinson to fuse the illumination of reason with the warmth of passion—and in the passage above one could add, with the energy of action.[23] In this poem the blurriness of those boundaries provides the author a distinct political advantage. "The Progress of Liberty" was serialized in the 1798 *Morning Post*; although that paper's audience had radical sympathies, the poem's project of defending the French Revolution was still by 1798 a difficult one. Robinson is willing to claim a connection between freedom and enlightenment, but she can hardly depict the Revolution as a triumphant assertion of equality on abstractly rational principles. By 1798 those principles had begun to look subversive and atheistical. She presents the Revolution instead as a physically and emotionally impoverished nation's desperate effort to join Britain in the sunlight of freedom. Her emphasis accordingly falls on benefits of freedom that middle-class Britons already felt they enjoyed—Protestant liberty of conscience, for instance, and economic prosperity—not on controversial questions like parliamentary reform. If in the course of this argument "liberty" loses some of its specifically political content and boils down to a physiological source of middle-class aspiration, that is all part of the rhetorical strategy.

The poem justifies the French Revolution especially by arguing that the *ancien régime* enslaved work that ought to have had the right to profit from its own exertion. In keeping with the governing solar metaphor, Robinson represents this slavery as a privation of light—for instance, in this depiction of pre-Revolutionary Italy:

Time *was*, and mem'ry sickens to retrace
The tablet fraught with wrongs, when seasons roll'd
O'er the small hut of lowly industry
In dim succession of eternal gloom;
Tho' rosy morn upon the eastern cliff
Burst wide her silver gates, and scatter'd round
A bright ethereal show'r! (*RP* 3:35)

Though the sun gives its vitalizing power, and calls the inhabitants of the hut to work, the correspondent light of liberty is missing. In fact, Robinson's solar metaphor allows her to glimpse an irony in agricultural toil that is not unlike Marx's early theory of alienated labor (with the difference, to be sure, that Marx blames private property for the alienation, whereas Robinson blames the French aristocracy for usurping property that ought to have been private).[24]

What was their hapless lot? To sigh, to pant,
To scorch and faint, while from the cloudless sky
The noon-tide beam shot downward, By their hands
The burning ploughshare thro' the Tuscan glebe
Pursued its sultry way: the smoking plains,
Refreshed by tepid showers, receiv'd the pledge
Of future luxury. The tangling vine,
Nurs'd by their toil, grew fibrous: the brown rind
Dried by the parching gale, wove close and firm,
Guarded the rich and nect'rous distillation.
. . . .
Yet for *them*
Did summer gild the plain? Did autumn glow? (*RP* 3:35)

The answer of course is "no." The beams of the sun that should represent human energy and work in concert with it, instead oppress the agricultural worker and ripen fruit destined for someone else's table. The arrival of French Revolutionary troops, by contrast, is described as a genuine sunrise— as blessed as the first dawn after the long "night of months" in northern Scandinavia (*RP* 3:33). The irony glimpsed here in Italy applies even more literally to the African slave. Robinson makes her abolitionist argument hinge on the oddly physical conceit that the southern races, "Born in the bland effulgence of broad day," ought to have more freedom than the inhabitants of the Arctic, where "despot darkness reigns, in sullen pride / Half the devoted year" (RP 3:29–30).

According to Robinson, the clerical and aristocratic castes were under the ancien régime as starved for light as the populace. A representative Monk, for instance, is "a tree / Plac'd on a blasted desert, where no sun / Visits the sapless trunk." Without the sun of liberty, there is no "electric spark" of fancy, no "perception strong," no "active mind," no "ambition," no "ethereal essence that expands the heart," only drudgery (RP 3:11). Without sensibility the body cannot perceive the world or initiate action; likewise, without liberty, the soul cannot act, but only mechanically repeat. Robinson is here drawing on old Protestant objections to Catholic ritual, but she folds those objections into her poem's larger naturalistic argument: political oppression excludes light, and thereby deprives bodies of their power of spontaneous feeling and action. The emphasis is less on the corruption of religious worship than on the waste of human potential: superstition is a "benumbing foe / To all the proudest energies of man!" (RP 3:27).

In depicting "the vestal sad" who is the Monk's female equivalent, the poem once again emphasizes the imprisonment that keeps her in perpetual gloom—in her case a literal confinement within the walls of a convent, where "fervid noon / No streamy light bestow'd to gild the cell." Here the discourse of sensibility shows what is—to Robinson's way of thinking—a particularly advantageous ability to include characteristically female powers and desires as forms of liberty. Robinson's portrait of the nun focuses less on frustrated ambition than on frustrated love; but the warmth of feeling is for Robinson an aspect of human energy as important as the light of intellect. In fact, even physical beauty seems to be an aspect of liberty coordinate with intellect and feeling.

> Was it for this her rosy infancy
> Was nurs'd with tender care? Her perfect form
> Fashion'd by all the graces and the loves,
> Rear'd to the op'ning summer of delight,
> A model of perfection? Was her mind,
> Stor'd with the prodigality of nature,
> Expanded, warm'd, enlighten'd, and inspir'd,
> For this to perish? Can the sable vest,
> The lawn transparent, or the pendent cross,
> Deceive th' omniscient! while her beating heart
> Proclaims her form'd for rational delight?
> Prepost'rous sacrifice! (RP 3:9)

Robinson views the nun's beauty as one more aspect of her potential fostered by nature and unjustly confined by the ancien régime. In "The

Progress of Liberty," in fact, all facets of nature's "prodigality"—all genial, glowing, and lustrous powers—seem to become aspects of liberty. This is what makes the solar metaphor so useful for Robinson, and distinguishes her poem from Thomson's. Where Thomson uses the sun's "forming light" to imply that liberty depends on a specific organization of society— especially, a distribution of rank and property—Robinson is interested in the sun simply as a synecdoche for nature's pervasive animating power. That metaphor gives her a way of arguing that England's collective political health depends on individuals' untrammeled pursuit of their own prosperity and pleasure—including, for instance, the pleasures of refined feeling and personal adornment. This is not quite the radicalism of Godwin or Wollstonecraft—who both insist that reason alone elevates a human being to dignity and freedom, and disparage "wealth and mere personal charms" as "unnatural distinctions established in society."[25] But it may be an even more characteristic form of the period's emerging middle-class consciousness. The furniture-makers and shopkeepers, printers and virtuosi who catered to Britain's growing market in luxury goods were not uniformly interested in the rational and ascetic conception of liberty promoted by someone like Wollstonecraft. But most of them could take pleasure in a conception of liberty as a "strong life-loving flame," rather like the power of life itself (RP 3:38). This kind of liberty does not depend on ownership of land, and it can be displayed even by people who are barred from active political life—notably women, but given the distribution of parliamentary seats, also the whole populations of Leeds and Manchester. This kind of liberty seemed to reveal itself in independent labor and Protestant worship, in beauty and fancy, love and ambition, and incidentally in "the electric fire, whose vivid glow / Illum'd the darken'd sense of Britain's bard [Milton] / With full Promethean blaze. . . . " (RP 3:17). The conflation of electricity and sunlight in this last figure would be an anachronism in the world of Greek myth, but Robinson is not primarily interested in Greek myth. She is pursuing an analogy between middle-class liberty and natural force, and in passing finds Prometheus's theft of fire from the sun a convenient illustration.

The Natural History of Ambition

Humphry Davy's applications of galvanic electricity made him the most eminent British chemist of the Regency period, and he was eventually created a baronet. But Davy was born in Penzance in relatively modest circumstances; his mother had been orphaned and adopted by a surgeon, and Davy was apprenticed to a surgeon, expecting a medical career. His notebooks from the 1790s mingle chemistry with Godwinian social ideas and philosophic poems.[26] Davy's early scientific papers and poems are particularly obsessed with the notion that the world's physical and mental phenomena all spring from the action of a single formless force. He gives that force a social interpretation that closely resembles the interpretation given in Robinson's "Progress of Liberty," but he makes the argument even more physical. For Davy, middle-class aspiration simply *is* natural force; it is scattered over the earth by the sun, and returns to the sun at death.

"The Life of the Spinosist," one of Davy's most interesting poems, traces what one might call an "ambition cycle" in nature. The poem was frequently revised, and has a complex history. The earliest recoverable copy appears in a notebook that Davy kept between 1799 and 1802; it was probably written late in 1800, since a letter dated October 9, 1800 from Coleridge to Davy refers to a slightly earlier version, now lost. Some of Coleridge's corrections to diction have already been incorporated in this notebook copy, although his main criticism, that "the thought, which you have expressed in the last stanza might be more gravely & therefore more consolingly, exemplified"[27] has apparently been ignored. In 1803 Davy attended a dinner with Coleridge and read a nearly identical version of the poem, transcribed unreliably by Clement Carlyon in his 1858 memoirs. Finally, in 1807, Davy substantially revised the poem, bending it in the direction of theism, and retitled it "Written After Recovery from a Dangerous Illness." This final version has been reprinted, but the earlier and, I think, more interesting poem has escaped notice.[28] The following discussion is based on the 1800 notebook version of the poem, with emendations in square brackets only where Carlyon's 1803 transcription fills a blank or corrects an obvious slip of the pen.[29] One erasure in the notebook is indicated with curved brackets.

In spite of its title, the poem shows little concern for Spinozist metaphysics as such. What passes for "Spinosism" in Davy's poem is late-eighteenth-century physics: instead of addressing issues of substance and being, the poem focuses on dynamics, describing the way life and motion are endlessly "renovated" in nature.

> Lo! oër the earth the kindling spirits pour
> The seeds of life that bounteous nature gives.
> The liquid dew becomes the rosy flower,
> The sordid dust awakes and moves and lives.
>
> All, all is change; the renovated forms
> Of ancient things arise and live again.
> The light of suns, the angry breath of storms,
> The everlasting motions of the main
>
> Are but the engines of that powerful will,
> The eternal link of thoughts where firm resolves

Have ever acted and are acting still,
While age round age and world round world revolves.

No sameness and no deep identity,
Linked to the whole, the human mind displays;
Impressible as is the moving sky,
And changeful as the surface of the seas.

Being of aggregate, the power of love
[Gives it] the joy of moments, bids it rise
In the wild forms of mortal things to move,
Fix'd to the earth below the eternal skies;

To breath[e] the ether, and to feel the forms
Of orbed beauty through its organs thrill;
To press the limbs of life with rapture warm,
And drink of transport from a living rill.

To view the heavens, with {solar} morning radiance bright,
Majestic mingling with the ocean blue.
To view the forests green, the mountains white,
The peopled plains of rich and varied hue.

To feel the social flame; to give to man
Ten thousand signs of burning energy;
The nothingness of human words to scan,
The nothingness of human things to fly.

To live in forests, mingled with [the whole]
Of nature's forms. To feel the breezes play
O'er the parched brow, to see the planets roll
O'er [his][30] grey head the life-diffusing ray.

To die in agony, in many days
To give to Nature all her stolen powers:
Etherial fire to feed the solar rays,
Etherial dew to feed the earth in showers.[31]

Like much of Davy's early poetry, "The Life of the Spinosist" includes a third-person figure who exemplifies the philosophic career that he sought for himself—a vaguely specified "Spinosist" who gives mankind "ten thousand signs of burning energy" before he flees "the nothingness of human things" for the forest. The poem is thus indirectly self-aggrandizing. But for a self-aggrandizing poem it is remarkably egoless. It is an understatement to call the poem's protagonist "vaguely specified": he enters the poem precisely as an *absence* of "deep identity" ascribed to the human mind, survives for several stanzas as the implied subject of infinitive verbs, and only emerges in his own right in the phrase "[his] grey head" before dissolving again into dew and sunlight. These instabilities suggest that the protagonist is standing less for Davy than for human life in general. Perhaps the best way to resolve the equivocation would be to say that the poem takes the arc of an ambitious young philosopher's career as a model of the source and fate of all human aspiration.

That source and that fate are described in distinctly physical terms. The poem begins with the sun's "kindling spirits," traces those spirits through a series of metamorphoses in nature and in human life, and ends by returning them, as "etherial fire," back to the sun. Plant life is produced by the action of light on "liquid dew"; animal life "awakes" and "moves" under the influence of the same light. These are not claims about spontaneous generation, but about the photochemical reactions that Davy, by 1800, understood as fundamentally sustaining animal and vegetable life. They parallel a more general metamorphosis that is perpetually occurring in the inanimate world. There is a significant order in the list, "The light of suns, the breath of angry storms, / The everlasting motions of the main." According to a meteorological theory Davy could have found in James Hutton's *Dissertations*, sunlight produces wind by warming air near the equator, and wind produces waves. These motions are thus all "renovated forms" of sunlight.[32] When the sequence of transformations reaches the human mind, the poem becomes hard for a modern reader to follow. The fundamental problem is that the poem fuses the concepts of stimulus and physical force, which are in the twenty-first century distinct. Guided here, as in his medical research, by a Brunonian rationale, Davy assumes that excitement—what we would call "stimulus"—is itself a kind of power. "Power" and "pleasure" are thus inter-convertible categories: when the "kindling spirits" of nature awaken the human mind, they do so as sexual pleasure ("the joy of moments"), or as the "transport" an infant drinks from the "living rill" of its mother's breast.

In short, pleasure plays the same role in the poem's account of human life that sunlight plays in its account of nature. Nor is it possible to keep the two distinct, for sunlight becomes a pleasure as well as a power when the protagonist views the "{solar} morning radiance" and the world on which that radiance confers color. The connection between light and pleasure was a matter of some importance to Davy.[33] In his 1799 essay on "Heat, Light, and the Combinations of Light," he had argued that the light contained in phosoxygen powers human consciousness, and that a deeper acquaintance with the chemistry of light might therefore make it possible "to destroy our pains and increase our pleasures" (*DW* 2:86). Sunlight was, for Davy and for many other writers in the 1790s, a privileged example of the link between pleasure and natural force. The mind gazing on the "life-diffusing ray" of the sky is, like the infant at its mother's breast, simultaneously drinking pleasure and vital power, excitement and nourishment, from a single source.

This analogy between the sun and the maternal breast forestalls the assumption that solar imagery is inherently male. Apollo the archer god is certainly a phallic figure. But the sun itself is frequently represented as female in the 1790s. Davy was consistently fascinated by a visual analogy between the sun and the "orbed beauty" of the breast; a number of other writers likewise represented the mind's dependence on natural force in terms of a child's relationship to its mother.[34] The prefatory poem to Erasmus Darwin's *Zoonomia* (1794) had already drawn a comparison between oxygen and mother's milk in language that Davy directly echoes: the embryo that "drinks with crimson gills the vital gas," grows into a suckling infant who "drinks warmth and fragrance from a living rill."[35] Breastfeeding is a central metaphor here, as in Davy's "Spinosist," because it can be used to show that human beings are surrounded, permeated, and kindled to consciousness by forces as alive and active as themselves.

For Davy, those forces were specifically associated with social aspiration. His poem "The Sons of Genius," written in 1795–1796 while he was still an apprentice in Cornwall, makes this quite explicit. The "sons of genius" are also "sons of nature"; they delight in "Nature's living fires," and, as in Davy's "Spinosist," they seem to drink pleasure and power from the sky. Their "minds / Imbibing portions of celestial day / Scorn all terrestrial cares, all mean designs" in order to seek "the glory of a lasting name." The light of "celestial day," in particular, becomes a figure for the spirit of activity that lifts certain people out of mean and provincial surroundings.

> While superstition rules the vulgar soul,
>   Forbids the energies of man to rise,
> Raised far above her low, her mean control,
>   Aspiring genius seeks her native skies.[36]

A modern paraphrase of the poem would have to turn the "sons of genius" into "philosophers" or "intellectuals," but for Davy this class of persons is consolidated less by a specific professional activity than by a special relationship to natural force and social aspiration.

Davy's "Spinosist," written when he was four years older, is somewhat less optimistic about society's receptiveness to aspiration. The opening lines of the eighth stanza ("To feel the social flame; to give to man / Ten thousand signs of burning energy . . . ") suggest that the pleasure the mind feels at being part of a social aggregate is transformed into productive energy. But the stanza registers a certain hesitation about returning energy to society; only "signs" of energy are given back before the protagonist scans the "nothingness of human words" and decides to flee the "nothingness of human things" in favor of philosophic retirement. The eighth stanza's hint that social pleasure returns to society as energy thus turns out to be abortive. But life and mental power do return to nature in the last stanza.

> To die in agony, in many days
> To give to Nature all her stolen powers:
> Etherial fire to feed the solar rays,
> Etherial dew to feed the earth in showers.

The stanza's unsparing first line is probably responsible for Coleridge's complaint that "the thought, which you have expressed in the last stanza might be more gravely & therefore more consolingly, exemplified."[37] (I am cynical enough to suspect that Coleridge really meant "more consolingly and therefore more gravely.") In understanding this last stanza it may help to know that Davy, like other philosophers of his time, believed that the phosphorescence of dead fish and decaying wood was a sign that all dead things gradually release the light they have absorbed in life: hence the gradual release of "etherial fire to feed the solar rays" (DW 2: 83–84).[38] This belief that the power of animation might linger for a while in apparently dead bodies led to Davy's lifelong fear that some residual sensation and intelligence might survive in his own body after death; for this reason, he requested that a period of ten days should elapse after his death before entombment—though Davy's brother was not, as it happened, able to fulfill the request.[39]

The crucial point about "The Life of the Spinosist," in sum, is its refusal to distinguish between the psychological, physiological, and social arcs traced by its protagonist's life. The nourishment the Spinosist drinks from his mother and from the sun is also pleasure ("rapture warm") and ambition ("burning energy"); the powers that return to nature on his death are intellectual powers, but also dew and light. Not every late-eighteenth-century poet was a chemist, but in conflating excitement and natural force, "The Life of the Spinosist" doesn't rely on especially esoteric knowledge. The crucial background for the poem lies not in Davy's specific theory of phosoxygen, but in two more general premises: first, the assumption that the sun's light somehow powers human life, and second, the assumption that pleasurable excitement is convertible with vital power. As chapter 1 suggested, both of these premises were widely shared in the 1790s: the first had been diffused by writers like Senebier, Berthollet and Lavoisier, and the second by Brunonian medicine. It should not be surprising, then, that early Romantic addresses to genius and to the power of poetry almost invariably represent that power as a sun.

Apollo and the Physiology of Liberty

In this chapter and in chapter 6, I propose that the Romantic cults of Prometheus and Apollo were aspects of a broader cult of autonomous enterprise, imagined as and represented as natural force. The poems of Humphry Davy and Mary Robinson support that argument, since they articulate their conceptions of poetic and philosophical genius through a larger analogy between middle-class autonomy and "Nature's living fires." Robinson's poems also hint that *Promethean* was in the 1790s a code-word for that analogy. But since neither writer has much to say about Apollo, they offer little direct support for the most controversial part of this chapter's thesis, which will presumably be the connection between middle-class work and Apollonian lyricism. Many scholars will already be willing to concede that Romantic interest in Prometheus had something to do with liberal politics, and perhaps even with "dynamic naturalism"; one does not have to press *Prometheus Unbound* very hard to see the convergence of those two themes. The figure of Apollo, on the other hand, has no obvious political resonance. It would appear much easier to explain his prominence in Romantic poetry as a reflection of writers' heightened ambitions for the lyric mode.

But perhaps critics take Apollo's relevance to poetic ambition too much for granted. Augustan poets talked about Apollo, made fun of him, and asked him to sting poetasters with his arrows, but they rarely invoked him to consecrate their own literary ambitions: that role was traditionally assigned to the Muses. Nor did they represent Apollo as an aspiring poet engaged in a project of self-creation (the role he plays in Keats's *Hyperion*, for instance). To understand how the Romantic Apollo became a god on the make, and a patron of poets on the make, one should begin by noticing that in the 1790s Apollo's inspiring power was not limited to poetry; it could become, like Robinson's "Liberty," a fiery and quasi-physiological power conferring initiative on all independent and enterprising persons.

Consider, for instance, the conflation of middle-class liberty and Apollonian fire in *The Helots*, a play by the Reverend Henry Boyd, published in Dublin in 1793. The play is based on an incident from Thucydides: Sparta is at war with Athens, and Sparta's Messenian slaves (the Helots of the title) have to decide whether to remain loyal, or to rebel and throw their lot in with the Athenians. They decide to remain loyal to their Lacedæmonian masters, which turns out to be a very bad move: the Spartans reward their misguided loyalty by inviting two thousand of their most enterprising young men into the Spartan army, in order to secretly massacre potential leaders of a revolution. The play was published in Dublin in the same year England went to war with France, and it could easily have been read as a despairing comment on Ireland's loyalty to the English crown in that struggle. This reading is not overtly supported by Boyd, whose preface suggests that the play is instead about "our negroes in the West Indies (the Helots of modern times)."[40] Announced authorial intention might not have governed the play's reception if it had been performed in Dublin, but that never seems to have happened. This lengthy drama based on Thucydides may never have been intended for the stage.

Boyd does acknowledge a parallel, not between Sparta and Britain, but between Sparta and prerevolutionary France. "There was no middle order in Lacedæmon, who might have preserved the balance between the two parties. All were imperious masters, or abject slaves, divided from each other, and kept apart by insuperable prejudices" (vii). This echoes, again, a contemporary discourse of complaint about the political consequences of Ireland's comparatively weak middle class. But rather than draw out this parallel, Boyd gestures to France, where the middle orders (by which he seems to mean both the lower gentry and the bourgeoisie) were decimated by the expulsion of the Huguenots and by strict adherence to primogeniture. *The Helots* is shaped throughout by the theory that these missing "middle orders" could have offered a way out of France's catastrophic opposition of "imperious master" and "abject slave." The tragedy of the play is, in effect, the Helots' failure to perceive and follow the alternative path of middle-class independence. This presents a technical problem: if there was no "middle order" in Lacedæmon (or in France), how can the play represent middle-class identity as if it had been a viable but tragically ignored alternative? Boyd solves the problem by allowing sun-gods to speak for the missing middle classes. The gods involved are, specifically, Apollo and the Persian deity Mithras. For the purposes of the play, they seem to be nearly interchangeable; each becomes a god of independence through his association with sunlight. The play's rationale for the connection between independence and sunlight is complex: it involves both an abstract argument about impartiality and a concrete physiological claim.

First Memnon—a Greek woman disguised as a Persian man, and incidentally named for a statue associated with the sun—gives the following account of the Persian deity Mithras:

> He that rules the day
> From his bright station deals impartial light,
> Both to the proud oppressor and the slave
> Who drags the clanking chain. The tyrant scorns
> Th' ætherial blessing, and the weary wretch
> But wakes to curse his rising beam, that shows
> A long variety of woe and pain.

But in the nightly visions of the just,
(After his radiant eyes have view'd the world,
Its miseries and wrongs) *he* deals around
That awful verdict oft, that seals the doom
Of thoughtless tyrants, tho' they bask secure
Beneath his blessed beam.[41]

The passage constructs a lopsided tripartite division of humanity into oppressors, slaves, and "the just." This looks like a case of crossed categories— since one would imagine that certain "slaves" might be included in "the just"—but the dilemma is resolved when the reader realizes that the just are standing in for the independent middle classes, unnameable in the play's Grecian context. Because they know neither scorn nor servitude, they are the only ones able to appreciate the "ætherial blessing" of sunlight. Their independence, in fact, allows them to share the sun's impartial perspective, and to become the divine instruments that work the downfall of tyrants.

The play's other connection between sunlight and independence is more tangible—indeed, more physiological—and is articulated by Apollo rather than Mithras. The Helots ask the oracle at Delphi whether the moment favors revolution, and Apollo responds without his usual ambiguity.

The only means to know
What moment favours freedom is to know
The time, when mortals dare to act or die,
When the existence of a slave is scorn'd
Compar'd with independence. Let them learn
(If not from men) from those proud savages
That roam the midnight groves, and thin the fold
With dark invasion—did they ever know
The trammels of a slave? or meanly fawn
For a poor pittance at a master's foot
Or draw the pond'rous plow? My instinct lives
In THEM. That Eleuthorian [sic] flame, that warm'd
The sons of Athens, when the Persian fled
Before his lifted spear! MY instinct lives
In every sinewy arm that wields the spade
Or goads the steer on yon Laconian plain
And would they learn my will, let them consult
The oracle within![42]

Apollo's answer, in effect, is that every moment favors revolution, because the instinct of independence is natural: it lives in every wolf, and in every Helot's "sinewy arm." The rationale for the claim is partly Protestant, as can be glimpsed when Apollo rather un-Delphically urges the Helots to consult their "oracle within." But *The Helots* transforms the old Protestant metaphor of "inner light" by making it strangely physical. The instinct of independence lives, not in the Helots' consciences, but in their sinews, and it is insistently fire-like. Here it is the "Eleuthorian flame" (the flame of freedom), but also, elsewhere in the play, "the glow / That lifts the slave to manhood," and "the flame of liberty." More than liberty is involved, moreover, when the Helots swear "by yon sun, / And all the powers that watch us as we soar / From slaves to manhood."[43] Apollo seems to preside over all masculine coming-of-age; the "flame of liberty" is also a flame of ambition.

Apollo and Poetic Ambition

In Greek mythology Apollo was not the god of poetry at all, but the protector of health and patron of physicians; his association with poetry evolved later as an accidental consequence of the fact that he was the owner of the first lyre.[44] By the seventeenth century, Apollo's connection to poetry was well established. But when Restoration poets wrote about Apollo, it was still most often in connection with the erotic myth of Daphne and Apollo, or in order to imitate Horace's ode to Apollo, in which, "He asks not riches of Apollo, but a sound mind in an healthful Body."[45] When Apollo did appear in connection with poetry, he had the specialized task of embodying masculine authority—for instance, he issued critical "edicts," or stung poetasters with his arrows. Ambitious poets still invoked the patronage of the Muses—especially Erato and Calliope—more often than they invoked Apollo. In the Romantic period, on the other hand, the invocation of the muse goes out of style, and Erato's name becomes a recondite allusion, while invocations of Apollo grow in number, seriousness, and fervency.

It is tempting to look for a psychosexual explanation of this shift, since a number of critics have argued that a remasculinization of the lyric genre was underway at the very end of the eighteenth century. But why would a remasculinization of poetic practice necessarily require writers to jettison female sources of inspiration? Male writers have often been quite pleased to dramatize their masculinity through a relationship to a feminine muse. Romantic writers were not shy about avowing dependence on the feminine principle of Nature; even the sun, as Davy has demonstrated, could be represented as a sustaining maternal breast. I am inclined to believe that the Romantic preference for Apollo over the Muses had less to do with Apollo's gender than with his connection to the sun—or in other words, with writers' desire to imagine literary production as a spontaneous and natural force.

Invocations of Apollo as the patron of poetic ambition seem to appear first as part of the cult of neglected genius. Elegies for "neglected genius"

and "neglected merit," popular from the latter half of the eighteenth
century through the Romantic period, allowed ambition to speak becom-
ingly in someone else's voice. Apollo appears in these poems, sometimes as
the patron of poets in particular, but quite often as the patron of neglected
merit in general. Anthony Pasquin composes a fable of "Merit and Envy"
(1787), for instance, in which the infant Merit is raised by Apollo. W. H.
Ireland, the engraver's son, admirer of Chatterton, and forger of invented
plays by Shakespeare, writes a whole volume entitled *Neglected Genius*
(1812), where Apollo's fires seem to emblematize the aspiring energies of
genius in general, and not just of poetry:

> I sing neglected worth, I mourn the doom
> Of genius slumb'ring in the silent tomb:
> I weep the sons of fire—Apollo's race,
> And blush to own my country's dire disgrace. . . .[46]

Much of the reason for Apollo's new prominence in this discourse is that he
can function simultaneously as a god of poetry and—in his capacity as sun
god—as a figure for the connection between ambition and natural force.
The two roles blend into each other, so that even in his role as god of
poetry, Apollo stands not so much for the particular genre of the lyric, as for
cultural aspiration in general. W. H. Ireland writes, for instance, in 1803,

> Arouse my soul, that I may snatch one ray,
> And claim alliance with the god of day;
> That I may wake from this lethargic dream,
> And own the influence of Apollo's beam;
> Then shall bright fancy spread her pinions wide,
> Then shall my genius dare some thought untried,
> And boldly soaring to bright realms unknown,
> Demand the new-found region for my own.
> Hark! 'tis Apollo tunes the dulcet lyre,
> I feel a spark of his celestial fire.[47]

Ireland isn't asking the god for guidance, but for a power of spontaneous
activity that will lift him out of his "lethargic dream." The invocation is thus
in effect an apostrophe to his own social ambition—as verbs like "claim
alliance" and "demand the new-found region" make only too clear. But by
invoking Apollo entirely in terms of sunlight ("celestial fire"), Ireland is able
to represent ambition as if it were a sacred thirst for a spontaneous power
that dwells in nature itself.

The equation between ambition and natural force was equally clear to
another admirer of Chatteron, when, after playfully crowning himself with

laurel at Leigh Hunt's house, he was stricken with seriousness, and composed "God of the Golden Bow" (1817) as a kind of retraction. The first two stanzas of John Keats's poem catalog mythological machinery, but the last stanza makes a more lasting impression, because the poem at last explains its reason for treating an antique god as if he were a living moral principle.

> The Pleiades were up,
>   Watching the silent air;
> The seeds and roots in Earth
>   Were swelling for summer fare;
>   The ocean, its neighbor,
>   Was at his old labor,
> When—who, who did dare
> To tie for a moment thy plant round his brow
>   And grin and look proudly
>   And blaspheme so loudly
> And live for that honor to stoop to thee now
>   O Delphic Apollo?[48]

This poem's tone is seriously abashed; one is reminded of Ian Jack's remark that "To Keats, Apollo was more than a figure of speech or a literary allusion: he was becoming something very close to the God of his adoration."[49] But the reason for Keats's remorse needs careful consideration. In 1817 he has no intention of renouncing poetic ambition. Nor has he blasphemed against Apollo by belonging to the lower middle class: the author is not that humble. Keats seems to feel that he has blasphemed by appropriating an outward sign of distinction (the laurel wreath), instead of relying on the inner force that is (in his view of the world) the only source of true distinction. That is why the most vivid contrast in the poem is the contrast between the crowned and grinning poet and the ubiquitous action of natural force (the roots swelling in the earth, and the ocean "at his old labor"). Poetic ambition is only sacred when it is like those things; since its claim to universality is founded on dynamic naturalism, its labors must be as ceaseless and spontaneous as the ocean's. Or, to put this in the more famous language of a letter composed the following year: "the rise, the progress, the setting of the imagery should like the Sun come natural natural too him. . . . if Poetry comes not as naturally as the Leaves to a tree it had better not come at all."[50]

Keats explored the aptness of Apollo's double role as sun god and god of poetry in a number of other poems composed between 1816 and 1818. On occasion he gives the connection between poetry and sunlight the same kind of physiological literalness this chapter has traced in Davy and Robinson. A poem enclosed in a letter to J. H. Reynolds, for instance, playfully discards "burgundy, claret and port" in favor of the poetic stimulant of sunlight: "My bowl is the sky / And I drink at my eye." In "Sleep and Poetry," the poet imagines himself worshipping the "wide heaven" of Poesy from a mountain-top, praying to "die a death / Of luxury, and my young spirit follow / The morning sun-beams to the great Apollo"—which is a somewhat lighter-hearted version of the solar reabsorption of ambition that takes place at the end of Davy's "Spinosist."[51] Keats does not, like Davy, explore the chemical connection between genius and sunlight. But it would be wrong to say that the association between sunlight and poetry in these poems is governed by the fact that two concepts happen to converge in the mythic personage of Apollo. Like Robinson's analogy between liberty and sunlight, Keats's analogy folds metonymy into metaphor, blurring the boundary between tenor and vehicle to achieve a characteristically Romantic literalness. Since poetry is one of the ways a fine day can animate the mind, the psychological effects of light are linked to poetry literally. But the physical effects of light are also compared to poetry figuratively: like light, poetry governs the world without any appearance of strain, so that its power almost seems coextensive with the effects it causes.

> A drainless shower
> Of light is poesy; 'tis the supreme of power;
> 'Tis might half slumbering on its own right arm,
> The very archings of her eye-lids charm
> A thousand willing agents to obey,
> And still she governs with the mildest sway. . . .[52]

This metaphor differs from the ancient metaphor of "inspiration" rather as Mary Robinson's theory of liberty differed from James Thomson's. When the muses inspired poets, they traditionally provided an inkling of a design, or at least the leading thread of a narrative; one asks the muse to "Tell me [specifically] of the man of many ways, who was driven far journeys, after he had sacked Troy's sacred citadel."[53] The Romantics' Apollo is in some ways a more accessible patron deity, since he becomes visible in the waves of the ocean and the growth of plants, not to mention the rays of the sun. But the inspiration he provides has neither form nor content to it: like Robinson's liberty, it is pure agency and pure pleasure, the "parent of delight" and "source refin'd / Of human energy" (RP 3:1).

The texts I have been considering are all located between 1816 and 1818; and in Keats's short career, the distance between the beginning and the end of 1818 is admittedly all the distance in the world. In the more famous works of 1819—Hyperion and the Odes—Keats's youthful analogies between poetry and natural force are complicated and transformed by a distinctly historical melancholy. The discussion of Byron and Shelley in chapter 6 will indirectly acknowledge some of the reasons for that darkening of tone.

Los and the Sun of Energy

William Blake enters this book not to provide evidence for its thesis but to explore its outer edges. Labor is a central theme in Blake's poems, and the sun figures importantly in his iconography, but in connecting the two he overtly resists the sort of connection I have traced in Davy, Robinson, Boyd, and the young Keats. The quickest way to get at this is to observe that though the name *Los* is *Sol* reversed, Los is not in the ordinary sense a sun-god. He is the creator of the sun. And though in some myths it might not make a great deal of difference whether you forge the sun, or draw it in your chariot, or embody it in your person—one metonym being just as good as the next—in Blake it definitely does matter whether the sun is understood as a portion of human existence or the other way around. In the illuminated books engraved between 1793 and 1795, the creation of the sun is part of a fall away from undivided humanity. It takes place only after Urizen has sundered himself from the other Eternals, thereby separating reason from other human faculties and desires. Light itself is a consequence of that primal secession, as the third chapter of *The Book of Los* (1795) explains. In gathering the "subtil particles" of light and forging them into an orb, Los restores a measure of order to the chaos left by Urizen's departure, and the sun may be in that sense (like Los's work in *Jerusalem*) a step toward the eventual reunion of human faculties and the redemption of man. But that horizon is still distant in *The Book of Los*. What Los more immediately achieves by creating the sun is the perfection of a "Human Illusion."

> 7. Nine ages completed their circles
> When Los heated the glowing mass, casting
> It down into the Deeps: the Deeps fled
> Away in redounding smoke; the Sun
> Stood self-balanc'd. And Los smild with joy.
> He the vast spine of Urizen siez'd
> And bound down to the glowing illusion
> 8. But no light, for the Deep fled away
> On all sides, and left an unform'd
> Dark vacuity . . . (*BP* 94)

The sun is a "glowing illusion" because it opens up an illusory space that seems to separate human subjectivity from the world, as Blake indicates here by letting the binding of the sun to Urizen's spine confirm the "unform'd / Dark vacuity" around him. In this cosmogony, the division of light from the space it illuminates parallels a broader contraction of consciousness, till at last human beings confine themselves to peering at the world through the five chinks of the senses.

Since Blake insisted on understanding the sun as a fallen portion of humanity, he was unwilling to idealize consciousness and freedom as Robinson, Boyd, and Davy did, by representing them as portions of an "etherial fire" borrowed from the solar orb. Like Davy, Blake argued that human consciousness was fundamentally one with the physical world, but he wanted to recognize that union by assimilating the world to human personality, and never by assimilating humanity to the abstraction "nature." His views on this question were clear, and did not change significantly between "There is No Natural Religion" (1788) and Jerusalem (1804–1809, published 1818). Very few British writers shared Blake's firmness of conviction on this point. Though many grappled with the same question, and some ended by hinting that nature and human imagination were two words for the same power, none insisted as clearly as Blake that the underlying power should be understood as an integrated human personality. Percy Bysshe Shelley, for instance, was an ontological idealist. But the thrust of his idealism was more to depersonalize mind than to personalize nature; his poems dazzlingly blend abstract imaginative power with unseen winds and voiceless lightning. The meditative dimension of William Wordsworth's poetry perhaps approaches more closely to Blake: certainly it foregrounds self-consciousness as a problem. But as Neil Hertz sensitively observes, Wordsworth's "heightened subjectivity" typically "creates not a more highly individualized subject, but a more impersonal and generalized one."[54] Wordsworth achieves this effect of impersonality—as I argue in the next chapter—in part through fantasies of mobility and ubiquity that bring him closer to scientific naturalism (and even to industrialism) than critics have usually acknowledged. It would be possible, then, to stop discussion of Blake here, allowing him to stand as a solitary exception to the period's tendency to idealize human agency by linking it to diffuse and impersonal force.

But although Blake is admittedly an exceptional figure, he is exceptional for reasons that run deeper than idiosyncrasy, and that have particular relevance to the present study. Blake attempted to connect an artisanal view of work—not otherwise well represented in Romantic poetry—to a prevailing idealization of work as spontaneous energy. The sparks that flew from that connection cast an interesting light on the tensions between different segments of middle-class opinion in the 1790s. I have argued that middle-class writers compared work to natural force in order to suggest that work carried with it a politically significant autonomy. Blake affirms this fusion of work and liberty as systematically as anyone in the period—embodying it in the figure of Los. And, like many other middle-class writers, he embodies it in solar iconography. But unlike most writers in the period, Blake refused to imagine energy as a freely circulating force. For reasons connected to his artisanal profession, he insisted that productive energy came indivisibly embodied in personality.

Blake's sun is more than a figure for the Human Illusion. In *The Book of Ahania* (1795), Fuzon (occupying Orc's usual role as embodiment of revolution and of fiery energy) molds a flaming globe and throws it through Urizen, thereby severing him from Ahania, "his parted soul." The globe is not quite the sun, but after it wanders the earth for 500 years as "a pillar of fire to Egypt," Los takes it up and forges it "in a mass / With the body of the sun" (*BP* 84–85, plate 1). One finds the same identification of revolutionary prophecy with "a pillar of fire," and the same implication that the sun is its product, in *The Song of Los* (1795) (*BP* 69, plate 8). The frontispiece to that book shows an aged man (perhaps Urizen) kneeling at an altar before a sun overwritten with spots or (in some copies) hieroglyphs—an "occulted sun" in Erdman's apt phrase, which acknowledges the text's reference to Urizen's religious mystery as "woven darkness."[55] Given the text's opening suggestion that Urizen has appropriated the prophetic children of Los as vehicles for his "Laws," one can infer that he has also appropriated the sun as an emblem of his religion of mystery. The final illustration in the volume shows Los resting on his hammer on a cloud, gazing down wearily or compassionately at a blood-red sun now free of writing. Erdman persuasively infers that "he has purged the sun of the runes of Urizenic mystery."[56] Alternatively, one could say that the runes have been purged by "the thick-flaming, thought-creating fires of Orc," the son of Los, since in this text Los is more a bardic observer than a participant. In either case, this sun, demystified but bloody, would seem to represent a state of consciousness immediately after revolution has disentangled prophecy from the webs of religion. In that capacity it could include *The Song of Los* itself, which would tend to identify the hammer of Los with Blake's own metalworking tools.

Since this sun of prophecy has a different valence from the "glowing illusion" to which Urizen is bound in *The Book of Los*, some readers of Blake (notably Northrop Frye) have been driven to the conclusion that "there are in Blake . . . two suns, one alive and unfallen, the source of life and light, and one dead, a 'phantasy of evil Man.' . . . "[57] I am not convinced that the early prophetic books require readers to draw such a systematic distinction. Though Blake is skeptical about the abstraction "nature," this skepticism does not necessarily, in his early works, imply outright hostility to natural forms.[58] Blake believed that the disk of the physical sun was an illusion, but

Figure 3.1   William Blake, *The Song of Los*, plate 1. Library of Congress. Page 126.

he chose to have this illusion forged by Los, the Eternal Prophet, rather than Urizen, the primeval Priest. The choice seems to suggest that the pleasures of light and vision are fallen and partial versions of what prophecy would be in Eternity. Though the bloody sun of revolutionary clarity that appears at the end of *The Song of Los* has been disencumbered of one layer of mystery, it remains fallen and bloody. There is thus no need to rigorously

Figure 3.2   William Blake, *The Song of Los*, plate 8. Library of Congress. Page 126.

distinguish it from the physical sun; both are imperfect glimpses or aspects of the same power.

One of the most interesting choices in Blake's writing is his decision to emblematize his own poetic and artistic work in the figure of Los. Los as Eternal Prophet is more independent of audience and patron than real writers and artists could ever be. But Los is also, as blacksmith, closely connected to toil, necessity, and constraint. Los undertakes his binding of Urizen as a desperate response to uncontrollable circumstance: "He watch'd in shuddring fear / The dark changes & bound every change / With rivets of iron & brass." This labor is compelled, and produces more compulsion; it is part of Urizen's fall, and not in an absolute sense a good thing. Blake nevertheless makes it seem heroic: "The Eternal Prophet heavd the dark bellows, / And turn'd restless the tongs; and the hammer / Incessant beat; forging chains new . . . " (BP 74–75, Urizen plates 8–10).

What is unusual about the passage is perhaps less its acceptance of the body than its acceptance of boredom. William Godwin, for instance, was quick to acknowledge the importance of physical exercise, but still indicted "mechanical and daily labor" as a pointless waste of human potential, because it suspends the free expression of individual energies. Blake can see something redemptive in the skill of the craftsman even when the task itself is painful or repetitious. Class obviously informs the difference between these two representations of work. Godwin was the son of a Dissenting minister, educated at a Dissenting academy; the members of his circle were teachers, natural philosophers, and writers. In thinking about work, he typically seeks to extend his class's vision of its own work as a freely chosen vocation, coextensive with republican and religious liberty. Blake was the son of a hosier, apprenticed early as an engraver. He locates the dignity of work less in absolute self-determination than in the skill possessed by the craftsman. Where Godwin tends to see labor as an evil to be eliminated in favor of free vocation, Blake is more immediately concerned with the contrast between skilled and unskilled labor. This concern emerges, for instance, in the well-known passage where "the Sons of Urizen" abandon "the plow & harrow, the loom / The hammer & the chisel, & the rule & compasses," as well as the hourglass and waterwheel,

> And in their stead, intricate wheels invented, wheel without wheel:
> To perplex youth in their outgoings, and to bind to labours in Albion
> Of day & night the myriads of eternity that they may grind
> And polish brass & iron hour after hour laborious task!
> Kept ignorant of its use, that they might spend the days of wisdom
> In sorrowful drudgery . . . (BP 216, Jerusalem plate 65)

Blake's "intricate wheels" enslave laborers above all by perplexing them; there is no way to acquire skill and rise within this craft. When, in the first book of Jerusalem, Los compels his Spectre to labor in the furnaces and build Golgonooza, there is just as much toil, compulsion, repetition, and lamentation (BP 152–55, plates 9–12). But there is not this perplexity. To be redemptive, apparently, work does not have to be spontaneous; but it does have to express the skill of the individual worker.

This emphasis certainly reflects the concerns of the artisanal lower middle classes, whose wages were at this time under pressure from engines that made it possible to replace skilled with unskilled labor. But as Morris Eaves has shown in his analysis of the "Public Address" Blake composed around 1810, it more immediately reflects Blake's own position in the art market. Journeymen had long been employed for the painting of draperies or the finishing of line engravings. In the late eighteenth and early nineteenth centuries, stipple and mezzotint techniques made it even easier to separate conception and design from the repetitive dotting or scraping of the plate that produced modulations of tone.[59] Though Blake himself used some of these techniques and though Catherine Blake did some of his coloring, his pronouncements about art attack styles based on modulations of tone ("Blots & Blurs") and insist on the importance of "Drawing" and "Line," which he identifies with invention (*BP* 575, 571; see also Eaves 178). More radically, Blake rejects all division of labor, and insists on the absolute inseparability of invention and execution. "Ideas cannot be Given but in their minutely Appropriate Words nor Can a Design be made without its minutely Appropriate Execution" (*BP* 576). Blake's underlying concern is that the cheap but laboriously finished work of journeymen will drive his own out of the market:

> the Lifes Labour of Ignorant Journeymen Suited to the Purposes of Commerce no doubt for Commerce Cannot endure Individual Merit its insatiable Maw must be fed by What all can do Equally well at least it is so in England as I have found to my Cost these Forty Years. (*BP* 573)

As Eaves shows, Blake's "Public Address" tends to indict all classes: the frivolous taste of aristocratic buyers, the cunning of middle-class print merchants who actively conspire to corrupt that taste, and the vulgarity and ignorance of the unskilled "hirelings" they employ. If Blake's defense of "Individual Merit" has something in common with middle-class rhetoric, it is perhaps even better understood as an attempt to be "all classes in one." It is, as Eaves precisely describes it, "a prospective expression of ambition, in which a utopian Christian vision legitimizes the desire to penetrate the confederation of devouring upper, controlling middle, and exploited lower classes at its mid-point to make economic room for autonomous middle-class artist-heroes who see the possibility of combining in themselves the aesthetic judgment of an elite and the technical skills of the working class."[60]

The labors of Los have the same breadth of social reference. Like other middle-class solar deities, Los embodies individual enterprise; since he creates his own furnaces and indeed his own body from scratch, he is in the most literal of senses a self-made man. Like Robinson's solar Liberty or

Ireland's Apollo, Los symbolically frees work from dependence on purchaser or patron—and in that sense reaches up to appropriate the economic independence of the gentry. But unlike most of the period's middle-class representations of work, Los also reaches down to appropriate an artisanal justification of labor as painful toil dignified by the personal skill of the laborer. As the "Public Address" makes clear, Blake's resistance to the division of labor in his own profession gradually evolved into a universal claim that true agency is never fungible, divisible, or transferable. Though Blake was attracted to the concept of energy, and though he used it like Humphry Davy to fuse body with soul, Davy's vision of energy as a fluid and ever-changing relational principle with "No sameness and no deep identity" would have been anathema to him. For all his criticism of selfishness and of spectral selfhood, Blake remains firmly attached to the principle that agency can only come in person-sized packages. When Blake writes, "Energy is the only life and is from the Body," he means from some *particular* body, not "from matter" or "from the forces of nature" (*BP* 34). There were religious reasons for this attachment to personhood, but equally fundamental was Blake's attachment to an artisanal vision of work, where labor is dignified by the strength and skill of the individual worker. Blake, however, makes the indivisibility of agency more absolute than it had ever really been in artisanal practice: the division of labor between master and journeyman, for instance, is sublated by the original artist who combines both roles.

Blake's contrasting example thus brings out some of the class interests left unstated in the dominant middle-class tradition. If Blake's commitment to inalienable skill qualified his interest in a connection between productive energy and natural force, it is fair to remark that other writers' unqualified enthusiasm for that connection takes the division of labor for granted. On the whole, Robinson, Davy, and Boyd had more to gain than to lose from the reorganization of labor; as an actor turned writer, a natural philosopher, and a Protestant minister, they belonged to occupations that were not yet in danger of being displaced by unskilled workers. The move to generalize from individual bodies to an agency that circulates through them and animates them posed no threat to their occupational identities, and might even have flattered them. In short, when I say that analogies between work and natural force were prevalent among middle-class writers in the 1790s, I must also acknowledge that different segments of the middle classes drew different lessons from that analogy. The project of describing work and liberty as coordinate forms of energy provided a necessary alternative to agrarian theories of independence and political virtue. But an artisan like Blake needed to qualify that project by insisting that energy was a personal possession. The implications that emerged when the analogy to natural force was taken wholly literally—particularly the implication that agency was at bottom impersonal and collective—seem to have had more appeal for commercial, clerical, and professional segments of the middle class.

In discussing middle-class conceptions of enterprise as natural force, I have tried to avoid a tone of heavy-handed indictment, because I assume that the distortions involved in that analogy will be tolerably plain to a twenty-first-century reader. Nature is not middle class, and middle-class production is not spontaneous. It depends on a network of social relations: on labor, on accumulated capital, and on a suitably organized market. Set next to this obvious idealization, Blake's older conception of work as the indivisible agency of an individual worker may appear refreshingly concrete. Critics of Blake have even suggested that his contrast between regulated and autonomous work anticipates the Marxian distinction between "alienated" and "unalienated" labor.[61]

That reading of Blake elides distinctions that are central to the present study. There is much that is admirable about Blake's determined effort to describe the world purely in terms of individual agency. It is related to the sinewy strength of his visual style, which often focuses on an individual body's relationship to a resisting environment.[62] It permits a critique of certain aspects of industrialization: particularly of the extreme division of labor in industry, which stifles human capacities and blocks off paths to occupational advancement. But a theory of alienated labor is exactly what Blake's system cannot produce. The Marxian theory of alienated labor depends first of all on the recognition that "the worker can create nothing without *nature*, without the *sensuous external world*. It is the material on which his labour is manifested, in which it is active, from which and by means of which it produces."[63] The laborer's necessary dependence on something outside human skill allows Marx to explain how labor, once externalized and stored up as private property, can return to tyrannize over the laborer. It also provides the normative force for Marx's critique of that tyranny. The laborer's dependence on products of labor alienated and appropriated by others is monstrous because it mocks, and estranges the laborer from, the fundamental dynamic of human existence—the original dependence on nature that constitutes "life-activity, productive life."[64]

In short, Marx's critique of alienated labor is not a critique of dependence as such: it requires a normative contrast between two different kinds of dependence. Blake, on the other hand, rejects the validity, and indeed possibility, of dependence on any world outside human consciousness. His denial runs deeper than his intellectual denial of a nature separate from man; it speaks impressively through his anthropocentric imagery. Blake consistently represents alienation as a fissure within the human body; Vala, or the veil of nature, is a personality emerging from one of those fissions. The subjugation of labor is thus always a direct subjugation of man by man.

This permits a vigorous political critique, but it collapses the space where economic critique could take shape, by denying the very possibility of dependence on an external world. That denial separates Blake, not just from Marx and Engels, but from an English tradition of social criticism running from Wordsworth, through John Ruskin, to William Morris—a tradition that (like Marx's own early work) used human interaction with nature to provide a normative contrast to social relations under industrialism. In rejecting other writers' idealization of work as natural force, and insisting that work is an indivisible personal possession, Blake did not escape middle-class ideology. He chose an older form of it, and in the process closed off certain critical opportunities.

Moreover, Blake's insistence that true agency must come in person-sized packages gave rise to its own kind of distortion. His penchant for interpreting the world in terms of the actions of individual agents made him the most persuasively mythopoeic writer of his time, but also contributed to a marked weakness for conspiracy theory. The "Public Address," for instance, blames the debased state of English taste on Louis XIV and on "Venetian picture traders" who set a program of "Counter Arts" in motion (*BP* 580). One is reminded of the disproportionately cosmic role personal enemies like John Scofield came to play in the later writings. Blake knew what he was doing: he was deliberately humanizing and individualizing a world that seemed to him to be headed in the opposite direction. His response is impressive and uniquely consistent, but it has, finally, some of the weaknesses of reaction. The impersonal solar energy and ambition depicted in Robinson's "Progress of Liberty" or Davy's "Life of the Spinosist" are less immediately satisfying than Blake's embodiment of labor in Los, because they frustrate readers' desire for a concrete human character. But the impersonality was making a necessary point. Without this late-Enlightenment insistence on situating human action within a larger system of agency, certain kinds of physical and economic insight would be impossible.

CHAPTER 4

COWPER'S SPONTANEOUS *TASK*

In the eighteenth century, Virgil's *Georgics* provided a model for poems that directly imitated Virgil's didactic manner—like John Dyer's *The Fleece*—and for poems that were "georgic" in the broader sense that they took agriculture as a subject. The georgic mode came to a stop, however, with or not long after William Cowper's long poem *The Task* (1785). Whether critics view *The Task* as a georgic poem, or as a heterogeneous collation of different modes including georgic and mock-georgic, they have agreed that it is hard to find significant georgic poems written after 1785.

For this fact, a variety of explanations have been offered.[1] But there may be less to explain than meets the eye. Georgic poems take the relation between human work and nature as their primary theme. In the eighteenth-century British georgic, however, the terms "nature" and "work" always tend to collapse into each other. Eighteenth-century georgic poets do not, like Virgil, exalt work as a conquest of nature; they exalt it as a natural process, remarking that "Even Nature lives by toil."[2] Instead of viewing *The Task* as the last gasp of an exhausted mode, one may as well view it as a consummation of this trajectory. Cowper finally succeeds in blurring the boundary between work and nature so fully as to create an intermediate category. Work is universally diffused throughout the physical world: "by ceaseless action all that is subsists." The restless "wandering" of the poem and its speaker are conflated with the spontaneous billowing of tree and cloud. By blurring this boundary, Cowper invents a kind of work that can be made to seem genteel. But he also erases the conflict that the georgic mode had originally taken as its subject.

Cowper's ability to make work genteel helps explain why *The Task*, as James King has remarked, "remained the most widely read poetical text in England until about 1800."[3] Cowper's popularity was especially high among the English middle classes, and lasted well into the next century. Francis Jeffrey wrote in 1811, for instance, "Cowper is, and is likely to continue, the most popular of all who have written for the present or the last generation."[4] Leonore Davidoff and Catherine Hall report that he remained by far the most widely quoted poet in middle-class memoirs and letters throughout the period between 1780 and 1850[5]—though this measure of popularity has its limits. (People who were reading Byron at night might well have chosen to quote Cowper in the letters they sent home.)

Some of the reasons for Cowper's popularity among English middle-class readers are not hard to guess. His fervent Evangelical religion certainly plays a role. But Cowper also gave expression to other kinds of ideals. I will argue that Cowper's primary achievement in *The Task* is to reconcile late-eighteenth-century idealization of work (which had social as well as religious roots) with the enduring notion that gentility precluded a life of mechanic toil.[6] Since many middle-class readers were similarly struggling to balance their aspirations to gentility with their pride in provident labor, Cowper's conflation of the two ideals found a ready market. Cowper locates a middle path between indolence and toil by folding his restless wandering into the inherent restlessness of nature. By comparison to this diffuse, spontaneous action, both indoor repose and the more mechanical forms of labor are made to appear artificial and (paradoxically enough) idle. *The Task's* sinuous rivers, ceaselessly swaying trees, and dappled plays of light illustrate this argument, making the woods and fields a visible emblem of the meandering, slightly unsettled form of work to which Cowper lays claim.

Since I am arguing that *The Task* reconciles two conflicting social and intellectual imperatives, the first part of this chapter briefly specifies what the "work ethic" and "ideal of leisured spontaneity" mean in eighteenth-century poems—and particularly in the georgic and descriptive poems that provided the most immediate models for *The Task*.

The Work Ethic in the British Georgic

The georgic mode was popular in the eighteenth century in large part because it allowed a sustained description of work; its practitioners tended to be Whigs who wanted to celebrate, rather than criticize, British prosperity.[7] But the representation of work in eighteenth-century georgic poems diverges significantly from classical tradition. Virgil's *Georgics* are not set in a pastoral golden age, but in the iron age, after the earth has ceased to give its sustenance freely. Jupiter, by inventing agriculture, has conferred on man a burden of labor, literalized above all in the act of plowing, "working the earth." The earth cooperates at most by providing the soil, sun, and rain necessary to accomplish the farmer's task. Left alone, the field will be "lazy," and plants will stop producing.

I have seen selected seeds, with care
Long tested, yet degenerate, unless
Man's effort picked the largest year by year.
So it is: for everything by nature's law
Tends to the worse, slips ever, backward, backward.[8]

In the classical georgic, agriculture is in fact primarily a battle or contest
between the power of man and the power of nature. "Toil mastered everything, relentless toil / And the pressure of pinching poverty."[9]

It is not very meaningful to talk about a "georgic mode" in English
before the eighteenth century, because didactic poems about agricultural
work are few and far between.[10] But the idea that labor is a uniquely human
burden had not disappeared with the advent of Christianity. Indeed, in the
case of such a good Protestant as John Milton, it sometimes invaded
surprising locales.

Man hath his daily work of body and mind
Appointed, which declares his Dignitie,
And the regard of Heav'n on all his waies;
While other Animals unactive range,
And of thir doings God takes no account.[11]

This is in Eden before the fall. But for Milton "the sweat of thy brow" is not
a curse; it dignifies human beings, and separates them from other living
things. The movements of animals are "unactive," by contrast, because they
are aimless.

It might be argued that poems like Ben Jonson's "To Penshurst" erase this
contrast, and represent land as doing work. Raymond Williams has pointed
out a peculiar "kind of wit" that represents the estate as spontaneously
productive, eliding human labor. Addressing Penshurst, Jonson writes:

Thy copse, too, named of Gamage, thou hast there,
That never fails to serve thee seasoned deer,
When thou wouldst feast or exercise thy friends.
The lower land, that to the river bends,
Thy sheep, thy bullocks, kine, and calves do feed;
The middle grounds thy mares and horses breed.[12]

Richard Feingold has identified similar rhetorical structures in Alexander
Pope's "Epistle to Burlington." ("Let his plantations stretch from down to
down / First shade a Country, and then raise a Town.")[13] This sort of rhetoric
is interesting because it replaces description of the social order with description of land, as Williams has shown.[14] But there is a significant gulf separating

this kind of wit from eighteenth-century postulations of a work principle in nature. To make land the subject of a sentence is to make it grammatically active, but it is not to suggest that land is laborious in a morally significant sense. Thus Jonson cannot, like Thomson, Dyer, and later Cowper, exhort his reader to follow nature's grand example of activity. In the estate-poems of both Jonson and Pope the moral example being held up is instead the example of the owner, who is implicitly praised for following principles of use (rather than show). Jonson expresses the praise metonymically, through a description of the estate itself, but the location of moral responsibility never becomes ambiguous—as it does in Cowper.

The tradition Cowper draws on is better understood as beginning with James Thomson's *Seasons* (1726–1746), which redefined the georgic mode for the eighteenth century.[15] Thomson echoes Virgil's *Georgics* explicitly in many passages, and they are the strongest classical models for the structure of the poem.[16] But Thomson transformed the georgic mode by making it descriptive rather than didactic. As Dwight Durling puts it, "Vergil painted the employments of the husbandman through the year, using the pageantry of the seasons as a background. Thomson described the seasons, using the employments of the farmer incidentally as part of his pageantry."[17] One reason for the altered perspective is that Thomson, as an early-eighteenth-century Whig from a middle-class background, is less interested in glorifying the self-sufficiency of the agricultural life than he is in celebrating a general union between British nature, British liberty, and British "arts" (taking that word to mean useful trades, and distinctly including the "sons of art" that fill Britannia's cities).

> Heavens! what a goodly prospect spreads around,
> Of hills, and dales, and woods, and lawns, and spires,
> And glittering towns, and gilded streams, till all
> The stretching landscape into smoke decays!
> Happy Britannia! where the Queen of Arts,
> Inspiring vigour, Liberty, abroad
> Walks unconfined even to thy farthest cots,
> And scatters plenty with unsparing hand.[18]

Thomson has little interest in using the life of the virtuous farmer, in Virgilian fashion, to criticize the luxurious city. On the contrary, the passage continues, "Full are thy cities with the sons of art / And trade and joy, in every busy street, mingling are heard. . . ." Britain's "glittering towns" are set next to its cots and "gilded streams" and blended into one hazy prospect of liberty and natural prosperity. In keeping with this shift in the georgic mode, the laborious protagonist of this poem is not so much the farmer in particular, as all of nature—or rather, God. But the distinction between nature and God has become, in *The Seasons*, a distinction that makes very little difference.

For instance, in "Spring," when the poem asks what "mighty breath" inspires love among creatures in that season, the answer comes in an assertion at once clear and self-effacing.

> What, but God?
> Inspiring God! who, boundless spirit all
> And unremitting energy, pervades,
> Adjusts, sustains, and agitates the whole.
> He ceaseless works alone, and yet alone
> Seems not to work; with such perfection framed
> Is this complex, stupendous scheme of things.[19]

The influence of physico-theology is evident here. It had long been orthodox to say that nothing happens except by the concourse of God. Newton's law of gravitation had given this old theological idea new vividness: gravity could now be understood as the immanent hand of God in nature, maintaining the planets in their courses. Thomson generalizes this idea to conclude that God is the sole and ceaseless worker in the natural order. But the natural order is "with such perfection framed" that God seems not to work. This is not deism; God is not absent from nature after framing it. Rather, God's "energy" (i.e., informing agency) in nature is so perfect, so spontaneously regular, as to efface itself. God's power is to be revealed not in miraculous intervention, but in nature, the miraculously lawful frame of his action.

The same sort of claim appears in John Dyer's didactic georgic about the British woolen trade, *The Fleece* (1757), but it is made directly as a claim about nature, not in the form of a claim about God. By 1757 Thomson's (and Newton's) elaborate way of attributing activity at once to God and to the natural order was beginning to collapse into a reference to "nature" alone. If in one sense this reduction to nature is a simplification, it also raises new rhetorical and moral problems that had been masked in Thomson's poem by the term "God." It's easy enough to understand in what sense God "works," but what moral significance does it have to say, with John Dyer, that

> Even Nature lives by toil:
> Beast, bird, air, fire, and rolling worlds,
> All live by action: nothing lies at rest,
> But death and ruin: man is born to care;
> Fashioned, improved by labor.[20]

If the planets "roll" by virtue of a *vis inertiae* and not by the immanent hand of God, it becomes difficult to understand how this activity can be described in any meaningful sense as "toil." The motion of the "rolling worlds" serves a valuable purpose, to be sure, but the worlds themselves make no exertion and overcome no resistance—two things that had been implicit in the definition of *toil* and *labor*.[21] Even the term *work* implied at least a distinct aim. Here there seems to be only endless circulation.

A similar contradiction, interestingly enough, appears in Smith's *Wealth of Nations* (1776). Smith argues that labor is the sole source of value, but also wants to represent agriculture as more productive than other trades. His solution is to argue that in agriculture, nature too counts as a productive laborer.[22] But if nature counts as a laborer, the statement that labor is the sole source of value would seem to lose its practical significance. Land is itself, effectively, a source of value, just as the physiocrats had claimed. This inconsistency within Smith's system opened a fracture that would eventually widen into a debate between David Ricardo and Thomas Malthus over the economic foundations of rent.

In short, Dyer faces the problem of reconciling a middle-class ethic that makes work the highest form of social activity with the eighteenth-century conviction that divine values are revealed in nature. This problem primarily troubled Whigs who were friendly to commerce (a group that included Thomson, Dyer, and Smith), and it became increasingly acute in the latter half of the eighteenth century.

Industrial Labor and the Georgic Mode

One obstacle to reconciling the georgic idealization of work with late-eighteenth-century naturalism lay in the changing character of work itself. Dyer's *Fleece* again provides succinct examples of the tension. The woolen trade that Dyer knew was no longer a trade that could plausibly be idealized through pastoral or georgic poetry. "Patient art," Dyer remarks, "has a spiral engine formed / Which on an hundred spoles, an hundred threads, / With one huge wheel, by lapse of water twines."[23] The poet then turns to address rural "nymphs," urging them not to fear that this "spiral engine" will cause technological unemployment.

> Fear not surcharge; your hands will ever find
> Ample employment. In the strife of trade,
> These curious instruments of speed obtain
> Various advantage, and the diligent
> Supply with exercise, as fountains sure,
> Which ever-gliding feed the flowery lawn.[24]

The simile in the last two lines deserves particular attention. Dyer is trying to find a poetic way of representing the idea that "spinning engines will merely provide more thread for weavers to weave." (Weaving was done on hand looms until the nineteenth century.) In keeping with the georgic mode he turns the thread into a never-failing fountain and the diligent weavers into a "flowery lawn." The industrial subject is thus transformed into an image of a securely cultivated estate. But—as Richard Feingold has remarked—the image fails to convince.[25] Dyer's figurative transformation of the factory system into cultivated land cannot connect with the old moral logic of the georgic mode: the ideal of farming as a life that is laborious, to be sure, but self-sufficient. Factory labor is a tragically inadequate basis for self-sufficiency—as Dyer's anxious apostrophe to his "nymphs" ("Fear not surcharge; your hands will ever find / Ample employment") makes plain without intending to.

Dyer's poem stumbles here, not because it "mystifies" industrial labor, but because it fails to do so. No poet had yet discovered, in 1757, how to successfully idealize industrial production. Classical poetic modes were inadequate to the task. This book traces the development of new ways of writing about work that did for industry what the georgic mode had done for agriculture. Dyer's celebration of industry contrasts interestingly, for instance, against Wordsworth's more effective celebration of industry in *The Excursion* (to be discussed in chapter 5). Even as he criticizes the factory system, Wordsworth will indirectly claim a place in nature for Dyer's spiral engines, by representing work as a natural force, and nature as an industrial laborer. In doing so he will also define a new kind of security (represented as physical mobility) that replaces the georgic ideal of self-sufficient cultivation.

William Cowper was not passionately concerned with technical advances in steam power, or with changes in the organization of manufacturing labor. This chapter examines the social origins of an analogy between work and natural force that writers like Wordsworth would apply to industrial labor several decades later. But there is little evidence that the analogy was at first fostered by industrialization. It began rather as an attempt to reconcile contradictions on the level of class. Those contradictions arose less from industrialization than from a commercial prosperity, founded on maritime trade, that had blurred boundaries between the landed gentry and the trading classes.

Spontaneity and Gentility in Cowper's *Task*

In the latter part of the eighteenth century the social function of nature was no longer primarily to distinguish the general and universal from "unnatural" aberration. On the contrary, natural behavior was often understood as behavior liberated from rules, given over to the individual's personal experience and idiosyncratic turn of mind. This shift comes through clearly in Cowper's pronouncements about *The Task*. In a letter to William Unwin accompanying the manuscript, Cowper links fidelity to nature and fidelity to the self.

> My descriptions are all from Nature. Not one of them second-handed. My delineations of the heart are from my own experience. Not one of them borrowed from books, or in the least degree, conjectural.[26]

In *The Task*, this fidelity to the self and to nature is also taken to imply spontaneity and irregularity on the level of form. Cowper chose to foreground the poem's adventitious origin in an "Advertisement" published with the poem.

> The history of the following production is briefly this:—A lady, fond of blank verse, demanded a poem of that kind from the author, and gave him the SOFA for a subject. He obeyed; and, having much leisure, connected another subject with it; and, pursuing the train of thought to which his situation and turn of mind led him, brought forth at length, instead of the trifle which he at first intended, a serious affair—a Volume! (*Task* 115)

This is first of all Cowper's way of explaining the poem's wandering structure and indeterminate genre. In "pursuing the train of thought to which his situation and turn of mind led him," Cowper produced a poem that begins with a mock-georgic account of the history of the sofa, and rises (through satire) to Christian revelation. Even Martin Priestman, whose thorough study of the poem's structure reveals a number of important organizing patterns, admits that the poem often "conducts an uneasy argument with itself," wandering between the genres of pastoral, satire, georgic, mock-georgic, and apocalyptic sermon.[27] Mental and physical wandering are in fact, as both Priestman and Dorothy Craven have observed, principles that structure the poem.[28]

The title of the poem remained *The Sofa* while the first four books were being written, but eventually changed to *The Task*, thereby centering attention on the whimsical "demand" from which the poem began. Since the end result is a poem in six books rising to the loftiest political and religious themes, the "task" has two senses, which undermine each other by turns: the playful assignment and the poet's chance-discovered "task" of serious composition, "a trifle" and "a serious affair—a Volume!" The juxtaposition of these ideas in the title is self-mocking, but its nonchalance also allows the writer to imply that he is above taking any "task" altogether seriously. The contradiction between the spontaneity that structures the poem and the labor in its title is also a social contradiction latent in late-eighteenth-century Britain.

For Cowper, this contradiction was embodied as a social divide. He was born to a Whig family in the minor aristocracy. After spending ten years in London, supposedly studying law, but primarily writing ironical essays and living as something of a rake, he found it necessary in 1763 to assume regular employment.[29] His uncle arranged an appointment as a clerk in the House of Lords, but the requirement to appear before the Bar of the House to defend his claim terrified Cowper. He fell into deep despondency and made several unsuccessful attempts at suicide as the date set for his examination approached, until his uncle finally released him from the appointment. He left London, spent a year in an asylum, and there became a pious Evangelical, to the horror of his family, since Evangelical Christianity was still largely a religion of the lower middle classes and the poor. Throughout the rest of his life Cowper suffered recurrent attacks of despair, and was filled with an increasingly acute conviction that he was irrevocably damned. He didn't return to London at all until 1792, and continued to feel that he was "unfitted . . . for society."[30]

Cowper's biography has been central to (and has even overwhelmed) criticism of his poetry, in large part because his life so well fits the conventions of his age. Representations of rural retirement in the late eighteenth century are frequently tinged with hints about persecution, ineradicable melancholy, and the necessity of fleeing to solitude. Cowper was not shy about taking advantage of this by dropping frequent allusions to his own inner torment in his poems, and the contrast between these allusions and the atmosphere of comfortable rural domesticity he creates has been a major source of his appeal to critics and readers. My reading of Cowper's poems need not deny the reality of his melancholy. But I am interested in the melancholy, not for its own sake, but for the way it reconciles a contradiction between Cowper's gentility and his Evangelical religion.

For Cowper, the ideal of natural spontaneity served, among other things, as a way of subtly laying claim to gentility. Although Cowper was not especially well-off, and seems to have been financially dependent on Mary Unwin when he wrote *The Task*, he remained quite proud of his gentle birth and continued to spend considerable sums on dress.[31] To return to the "Advertisement" to the poem: "I obeyed; and, having much leisure, connected another subject with it. . . ." The reference to "leisure" may (if taken in the larger context of the poem) allude to the melancholy that has driven the speaker into retirement, but it also clearly lays claim to gentility. The whole phrase borrows an aristocratic style of chivalry, representing the speaker as someone who "obeys" his lady friend precisely because he need not obey anyone else. Rural retirement, a central subject of *The Task*, is thus linked to the ideal of spontaneity that guides its wandering structure through the implicit gentility of leisure.

But spontaneity also had more serious connotations for Cowper. For instance, the "Advertisement" casually remarks that Ann Austen was "fond of blank verse," and demanded a poem in that form. Priestman has pointed out that blank verse had for Cowper an almost mystical significance drawn from its association with Edward Young, with Milton, and with a Protestant tradition of "Enthusiasm."[32] This high estimation of blank verse was part of Cowper's broader search for a "familiar style," a poetic language "Elegant as simplicity, and warm / As exstasy, unmanacl'd by form."[33] Even Cowper's pentameter couplets, such as the one just quoted, often use enjambments to "infus[e] . . . one line into another" instead of end-stopping the lines.[34] All of these choices resist the preferences of Tories like Pope and Samuel Johnson. But they also indicate a style of Whiggish enthusiasm that has learned to aspire not just to virtue, liberty, and prosperity, but to spontaneous elegance as well.

Cowper aimed for the same sort of effect in the organization of *The Task*. Each subject is to flow directly from the preceding one, as if they were themselves enjambed. In a letter to William Unwin, he remarks, "if the work cannot boast a regular plan (in which respect however I do not think it altogether indefensible) it yet may boast, that the reflections are naturally suggested always by the preceding passage. . . ."[35] I want to set the defensive parenthetical comment to one side for now. The main claim being made here depends on a distinction between formal or "regular" organization and the plan of organization that would spring "naturally" from the mind's associative processes.[36] The twisting path that results is compared explicitly and more than once in the poem to the winding, rambling path of the rural walks the poem spends so much time describing; the connection implied by the adverb "naturally" is thus made literal.

There was, in short, a strong connection in the author's mind between the poem's form—its blank verse, its serendipitously wandering structure— and the thematic "tendency" he claimed for it: "To discountenance the modern enthusiasm for a London Life, and to recommend rural ease and leisure as friendly to the cause of piety and virtue."[37] A social divide might seem to separate earnestly Evangelical "piety and virtue" from genteel "ease and leisure," but Cowper reconciles these social conflicts on the level of form. "Rural ease and leisure" are associated with the poem's naturally wandering organization; piety and virtue are associated with blank verse. God and unrhymed iambic pentameter become, through their connection to leisured spontaneity, ideals with a gentility of tone—in no sense the obsessions of a rustic enthusiast or a city drudge.

The poet's Evangelical faith tends, however, to undercut this resolution. Cowper's descriptions of rural ease are often shadowed by feelings of melancholy and even perdition. He expresses guilt especially by fretting about the difference between indolence and toil. When melancholy enters the poem, the narrator is often struggling to convince readers that his rural employments are "not slothful," but an oxymoronic kind of "laborious ease" (*Task* 3:361–62). This thematic anxiety is continuous with authorial anxiety about form: the lax structure of the poem might seem to imply an ungodly indolence. (Recall the defensive parenthetical remark I set aside earlier: Cowper's hope that "in respect of" a regular plan the poem is not "altogether indefensible.") Both author and narrator are concerned to make it clear that the poem is devoted to "No loose or wanton, though a wand'ring, muse, / And constant occupation without care" (*Task* 3:692–93). "Wand'ring," in particular, is a key word here; it refers at once to the inevitable straying of sinful man and to the poem's form. But, above all, it refers to an attempt to reconcile leisure and toil in the form of a restlessness at once rural and melancholic.

The Wandering Muse

Cowper's rural melancholy can be understood as a personal solution to a social problem: how to reconcile a Christian ethic of work and moral struggle with the late-eighteenth-century ideal of natural spontaneity. As a physical activity, vigorous but unconstrained, walking occupies a healthy middle ground between indolence and toil. But Protestant piety enjoined work for moral and not just for medical reasons. The demand for moral struggle, as much as any biographical trauma, determines the faintly unsettled or melancholy coloring of Cowper's rambles. The occasional, tantalizing reflections on an unnamed moral burden that tinge those wanderings become a way of implying (very "gently," in every sense of that word) that wandering belongs to a Protestant tradition of struggle with one's own conscience. None of this, I should emphasize, requires reading Cowper as a feigner, or denying that he really thought he was damned. To say that experience is constructed socially is not to deny its personal authenticity.

Book Three begins with a passage that compares the poem's meandering associative structure to a rural walk. The comparison is apologetic.

> As one who long in thickets and in brakes
> Entangled, winds now this way and now that
> His devious course uncertain, seeking home;
> Or having long in miry ways been foiled
> And sore discomfited, from slough to slough

Plunging, and half despairing of escape,
If chance at length he find a green-swerd smooth
And faithful to the foot, his spirits rise,
He chirrups brisk his ear-erecting steed,
And winds his way with pleasure and with ease;
So I, designing other themes, and call'd
T' adorn the Sofa with eulogium due,
To tell its slumbers and to paint its dreams,
Have rambled wide. In country, city, seat
Of academic fame (howe'er deserved)
Long held, and scarcely disengaged at last.
But now with pleasant pace, a cleanlier road
I mean to tread. I feel myself at large,
Courageous, and refresh'd for future toil,
If toil await me, or if dangers new. (*Task* 3:1–20)

The first three lines of the passage begin to tell a story about a traveler lost in thickets, winding on a "devious road uncertain"; the reader expects that this is an apology for having digressed from the poem's initial topic (the sofa), and therefore begins to expect a conclusion of the form "at last finds the straight well-traveled road." But in line four there is a syntactic break, and the emphasis shifts to the obstacles, pains, and troubles that have "foiled, and sore discomfited," the traveler. The apology seems to have shifted into a complaint; wandering has become trouble and toil. There are indications that the toil involved is at least partly a Protestant moral struggle: "from slough to slough / Plunging, and half despairing of escape" could easily evoke the Slough of Despond.

The next four lines conclude this latter story, not the first one. The traveler who has found the green-swerd has not found a straight road; he still "winds his way," but now he winds "with pleasure and with ease." The implication is that his earlier wandering constituted a form of un-ease, work. All of this has so far formed the vehicle of a simile; with "So I," the passage turns to explain the tenor. The comparison is difficult to establish with clarity, because it's difficult to be sure whether adorning "the Sofa with eulogium due" is a form of toil or a form of indolence. The mock-seriousness in this phrase corresponds to the equivocation in the poem's title. If "the Sofa" is a task to which the poet feels seriously "call'd," then his formal rambles become "wandering" in the sense of negligent straying (repeating the story with which the passage seemed to begin). If, on the other hand, the initial project of telling slumbers and painting dreams was a trifle, it would be the wandering itself that represented toil—virtuous at least to the extent that it constituted a moral struggle with despair. The dialectic between indolence and moral seriousness is clear, but the poem's meandering structure apparently falls on both sides of that opposition.

After a jeremiad against the pleasure-worship, slackness, and fraud of English society, the narrator suddenly pauses and begins an account of his own history. Here the equation of wandering with moral toil is made clearer.

I was a stricken deer that left the herd
Long since; with many an arrow deep infixt
My panting side was charged when I withdrew
To seek a tranquil death in distant shades.
There was I found by one who had himself
Been hurt by th'archers. In his side he bore
And in his hands and feet the cruel scars.
With gentle force soliciting the darts
He drew them forth, and heal'd and bade me live.
Since then, with few associates, in remote
And silent woods I wander, far from those
My former partners of the peopled scene,
With few associates, and not wishing more.
Here much I ruminate, as much I may,
With other views of men and manners now
Than once, and others of a life to come.
I see that all are wand'rers, gone astray
Each in his own delusions . . . (*Task* 3:108–25)

This passage recounts Cowper's breakdown, and his religious response to that breakdown, which was to retreat from London to the country. "Wandering" is, first of all, Cowper's way of attributing meditative serious-ness to this rural retreat."[I]n remote / And silent woods I wander. . . ."A sec-ond meaning of the word is implicit in the line "Here much I ruminate, as much I may," which links the poem's wandering structure to the leisure that rural retreat makes possible. This echoes the "Advertisement"'s way of explaining the poem's digressive structure: "He obeyed; and, *having much leisure*, connected another subject with it . . ." (my italics). But lest the reader begin to suspect that the contemplative freedom to "ruminate" is a form of indolence, the poem moves on to suggest that, in a third sense, wandering is nothing other than the fallen state of man itself. "I see that all are wand'rers, gone astray. . . ." Wandering can become a form of toil precisely because it is darkened by this consciousness of guilt—justified, as it were, by sin. It is not merely an indolent stroll, but a representation of man's spiritual task. By contrast, the ridiculously "industrious" philosopher who spends "the little wick of life's poor shallow lamp / In playing tricks with nature" lacks both moral seriousness and a genuine connection to the natural order (*Task* 3:161–66).

The dialectic between labor and indolence that has been structuring Book Three begins to resolve into a principle of "laborious ease."

> How various his employments, whom the world
> Calls idle, and who justly in return
> Esteems that busy world an idler too!
> Friends, books, a garden, and perhaps his pen,
> Delightful industry enjoyed at home,
> And nature in her cultivated trim
> Dressed to his taste, inviting him abroad—
> Can he want occupation who has these?
> Will he be idle who has much t'enjoy?
> Me therefore, studious of laborious ease . . . (*Task* 3:352–61)

Priestman has drawn attention to the odd alternation of "confession" and "self-vindication" in this passage, not just from line to line, but from phrase to phrase: "delightful industry," for instance, "enjoyed at home."[38] Cowper is driven to these lengths of defensiveness because he finds himself unable to justify retirement except as spontaneous toil. He cannot shun the busy world by appealing to a tradition of Horatian *otium* or, for that matter, a tradition of Christian meditation; he has to esteem the "busy world an idler," and make himself a student of "laborious ease." The labor of retirement is partly contemplative: "He that attends to his interior self / . . . / finds himself engaged t'atchieve / No unimportant, though a silent task" (*Task* 3:373–78). But it also involves writing, walking, and gardening—the last of these famously explored in the poem's long mock-georgic description of cucumber-frames. In all these domains, Cowper insists on the real toilsomeness of his apparent ease—distinguishing it all the while, however, from the sort of "lubbard labor" that would require "sinews bred to toil," which he assigns to servants (*Task* 3:400, 3:405). Gardening, like poetry, is devoted to "no loose or wanton, though a wand'ring muse / And constant occupation without care."[39]

"By Ceaseless Action All That is Subsists"

So far I have traced only one half of Cowper's argument about work: his attempt to find a middle ground between idleness and labor in melancholy wandering and formal digression. But while *The Task* reduces work to restlessness, it also elevates nature's activity to a kind of spontaneous labor. This aspect of the poem emerges most clearly in Book One, which begins with a mock-georgic account of the rise of seating, from "the rugged rock" through joint-stools and elbow-chairs. "So slow / The growth of what is excellent; so hard / T' attain perfection in this nether world" (*Task* 1:183–85). The humor here depends on a tension between the georgic celebration of invention and improvement, and the unlaborious subject of seating. Just when the zenith represented by the sofa itself has been attained, and the poem seems about to come to a rest, it leaps up from its subject in disgust.

Oh may I live exempted (while I live
Guiltless of pampered appetite obscene)
From pangs arthritic that infest the toe
Of libertine excess. The SOFA suits
The gouty limb, 'tis true; but gouty limb,
Though on a SOFA, may I never feel:
For I have loved the rural walk through lanes
Of grassy swarth, close cropt by nibbling sheep . . . (*Task* 103–10)

In the ensuing description of his boyhood walks the narrator does everything he can to undo the suspicion the introduction encouraged: that he is a luxurious idler. "I fed on scarlet hips and stony haws," he reports—"Hard fare!" He retains a youthful elasticity of foot and lung "that makes / Swift pace or steep ascent no toil to me" (*Task* 1:138–39). The words "no toil" may seem a strange way of making this point, but one should remember that Cowper does not want to claim that he actually *is* a laborer. The point is rather that he possesses a spontaneous activity for which toil is "no toil."

The poem moves into a description of a specific walk near the river Ouse with a "dear companion." Ostensibly the narrator forgets himself in admiration of nature. But descriptive details are also chosen to redefine the speaker. He does not actually labor, but acquires a vicarious activity by association with the activity of nature. For instance, "How oft upon yon eminence our pace / Has slackened to a pause, and we have borne / The ruffling wind, scarce conscious that it blew" (*Task* 1:154–56). A "ruffling wind" is not exactly something that has to be "borne," but the verb serves to suggest that there is a fortitude in the indifferent half-consciousness of poet and companion as the gentle motion of hair and clothing makes them one with the scene. "Ruffling" is characteristic of a whole set of words denoting gentle undulation that Cowper uses to project labor into nature. For instance, on this same eminence the poet sees two things: a tiny ploughman in the distance, and the river Ouse. They are described so as to assimilate the ploughman's labor to the winding of the river.

Thence with what pleasure have we just discerned
The distant plough slow moving, and beside
His lab'ring team, that swerv'd not from the track,
The sturdy swain diminish'd to a boy.
Here Ouse, slow winding through a level plain

> Of spacious meads with cattle sprinkled o'er
> Conducts the eye along his sinuous course
> Delighted. (*Task* 1:159–66)

If the plough, "slow moving," has its team and swain, so the Ouse, "slow winding," must have its cattle and its "eye" to conduct. The emphasis on the slowness produced by distance in both river and swain ends by softening labor to a sort of labor effect in the landscape.

The same impression is produced by the description of sound in this passage. The sound of distant bells "just undulates upon the list'ning ear," for instance. The next verse paragraph is more striking.

> Nor rural sights alone, but rural sounds,
> Exhilarate the spirit, and restore
> The tone of languid nature. Mighty winds,
> That sweep the skirt of some far-spreading wood
> Of ancient growth, make music not unlike
> The dash of ocean on his winding shore,
> And lull the spirit while they fill the mind,
> Unnumber'd branches waving in the blast,
> And all their leaves fast flutt'ring, all at once. (*Task* 1:181–89)

The last line defamiliarizes the image of the windswept edge of a wood by bringing the individual leaves "fast flutt'ring, all at once" into the foreground. The motion fills the visual field as the sound "fill[s] the mind." Other descriptive elements in the passage underline the same effect: nature is full of myriad gentle stimuli that repeat themselves rhythmically. The winds that "sweep the skirt" of the wood produce a kind of billowing that echoes the "winding shore" on which the ocean dashes. These stimuli are understood to "restore / The tone of languid nature"—the elasticity, in other words, of body and mind—in the same way it could be restored by gentle rhythmic exercise, like a rural walk.[40]

Cowper's way of laying claim to "laborious ease" is to argue that any immersion in nature (retirement to the country or, on a smaller scale, a rural walk) is by definition a kind of action or labor—because nature, in undulant motion, is active through and through. Nature is never idle. Therefore I, in nature, am never idle. This claim will shortly be stated in a general philosophical terms. But the poem first prepares the reader by embodying the claim descriptively. The speaker and his companion walk through an avenue with a swaying branchy roof,

> while beneath,
> The chequer'd earth seems restless as a flood

Brush'd by the wind. So sportive is the light
Shot through the boughs, it dances as they dance,
Shadow and sunshine intermingling quick,
And dark'ning and enlight'ning, as the leaves
Play wanton, ev'ry moment, ev'ry spot.
And now, with nerves new-braced and spirits cheered . . . (*Task* 1:343–50)

"Ev'ry moment, ev'ry spot" echoes "all at once" in the previous passage and
suggests the same slightly dazed fascination with the idea that change can
be going on *everywhere* in the visual field simultaneously. (Note that again,
this ubiquitous motion is represented as having a "bracing" effect on
observers.) The splotches of sunlight that swim on the ground make the
earth itself seem "restless as a flood" and thus begin to transfer the quality of
ubiquitous motion from a specific scene to the terrestrial world as a whole.

Now speaker and companion come on another emblematic figure
of labor.

                    The grove receives us next;
        Between the upright shafts of whose tall elms
        We may discern the thresher at his task.
        Thump after thump resounds the constant flail
        That seems to swing uncertain, and yet falls
        Full on the destin'd ear. Wide flies the chaff;
        The rustling straw sends up a frequent mist
        Of atoms sparkling in the noonday beam.
        Come hither, ye that press your beds
        And sleep not; see him sweating o'er his bread
        Before he eats it. 'Tis the primal curse,
        But soften'd into mercy; made the pledge
        Of cheerful days, and nights without a groan (*Task* 1:354–66)

The figure of the thresher is explicitly framed by trunks of elms, which help
to suggest that his labor is part of a "landscape"—understood here, as it
usually is in the eighteenth century, as a pictorial concept. The cloud of
chaff, "a frequent mist / Of atoms sparkling in the noonday beam," echoes
the checkered play of light and shade that was described a few lines earlier,
further assimilating the labor of threshing to the general restlessness of
nature. The climactic outburst that follows only generalizes the claim
already implicit in this image.

        By ceaseless action all that is subsists.
        Constant rotation of th' unwearied wheel
        That nature rides upon maintains her health,

Her beauty, her fertility. She dreads
An instant's pause, and lives but while she moves.
Its own revolvency upholds the world.
Winds from all quarters agitate the air
And fit the limpid element for use,
Else noxious: oceans, rivers, lakes, and streams,
All feel the fresh'ning impulse, and are cleans'd
By restless undulation. (*Task* 1:367–84)

This "ceaseless action" in nature justifies the thresher's labor; all of nature seems to be laboring beside him.

In these lines, Cowper transforms the character of the imperative to work. He referred a few lines earlier to the softening of a "primal curse," but in truth he has softened Adam's curse so much that it becomes unrecognizable. Work that is revealed in nature's "beauty" and "fertility"—in checkered light and shade, and wind sweeping the margin of a wood—is too Edenic to be understood as a curse. Nor does Cowper define work (like Milton) as an honorable badge of obedience to God. The very notion of obedience has been suppressed—displaced into the mock-heroic frame that makes the poem an assigned "task" of composition about the sofa. Serious work, by contrast, is here as spontaneous as the "restless undulation" of nature. All that is left of the Fall of Man is a very faint shadow of sin, or self-consciousness, that hovers around the word "restless." Nature feels this restlessness as much as man: "She dreads / An instant's pause, and lives but while she moves." Except for that faint shadow of dread, the poem has effaced the connection between work and compulsion, making it possible to imagine work (of a certain sort) as an expression of gentility.

Cowper wasn't the only writer in the late eighteenth century fascinated with wandering, or with undulant and ubiquitous motion. A few years later William Gilpin, for instance, extended the picturesque aesthetic to encompass a horse's glossy coat.

Such a play of muscles appears, every where, through the fineness of his skin, gently swelling, and sinking into each other—he is all over so *lubricus aspici*, the reflections of light are so continually shifting upon him, and playing into each other, that the eye never considers the smoothness of the surface, but is amused with gliding up, and down, among these endless transitions. . . .[41]

Gilpin foregrounds the simultaneous movement of innumerable separate parts—the muscles in motion "every where" and "all over." His fascination with this effect echoes Cowper's "Unnumber'd branches waving in the blast, / And all their leaves fast flutt'ring, all at once." Gilpin also seems to have interpreted images of ubiquitous motion as intimations of a work principle latent in nature. Roughness and activity are two of the chief principles of the picturesque. Gilpin likes them all the better when they are combined—in a ruffled surface or a ruggedly active form. For "altho the human form, in a quiescent state, is thus beautiful; yet the more it's [sic] *smooth surface is ruffled*, if I may so speak, the more picturesque it appears."[42] The transformation of labor into an aesthetic ruffling of the body's surface parallels the "labor effect" that becomes visible in Cowper's landscapes. Anne Wallace and Thomas Pfau have both suggested that the picturesque tour translated aristocratic appropriation of landscape into a form of virtual appropriation that could be practiced by a middle-class audience. Wallace is interested in the way pedestrian tours, in particular, fused a pictorial relationship to landscape with virtuous quasi-georgic activity. Pfau traces a connection between the "roughness" of picturesque scenery and the "spontaneity" of the middle-class viewer's aesthetic response.[43] I would confirm these arguments, and extend them slightly, by pointing out that there is also latent in Gilpin (as in Cowper) a scientific analogy between the ceaseless activity of nature and the spontaneous labor of the viewing mind.

Samuel Taylor Coleridge felt a similar fascination with images of ubiquitous undulant motion. In notebook entries he describes "Starlings in vast Flights, borne along like smoke, mist" and "Fern, on a hill side, scattered thick but growing single—and all shaking themselves in the wind."[44] In "This Lime-Tree Bower My Prison," he pauses to imagine the "long lank weeds, / That all at once (a most fantastic sight!) / Still nod and drip beneath the dripping edge / Of the blue clay-stone." The dialectic of stillness and motion is a complex subject in Coleridge's poetry. Stillness evokes a wide range of fears—fears of isolation, spiritual aridity, and unreality—and renewed motion addresses those anxieties in a variety of ways. But there is some evidence (at least in "This Lime-Tree Bower My Prison") that the fear of stillness is a fear of indolence. Images of ubiquitous motion seem to reassure the speaker in that poem partly because they convince him—as they convinced Cowper—that nothing is ever really idle. "No plot so narrow, be but Nature there, / No waste so vacant, but may well employ / Each faculty of sense, and keep the heart / Awake. . . ."[45]

In short, Cowper was not alone in seeking to acquire a patina of Protestant virtue through association with the "ceaseless action" of nature. But the social dialectic between gentility and labor that led Cowper to frame an ideal of spontaneous work is admittedly harder to trace in Coleridge's writings. The form of the conversation poem encourages a different kind of inner struggle, and Coleridge's explicitly political commitments drown out the subtler social anxieties a reader overhears in Cowper. For a clearer parallel to *The Task*, I turn in Chapter 5 to Wordsworth's *Excursion*, where a work principle diffused through nature allows the author to reconcile social tensions very like those Cowper struggled with.

CHAPTER 5

WORDSWORTH AND THE HOMELESSNESS
OF ENGINES

William Wordsworth's response to industrialization fused two conflicting impulses. His assessment of the machines themselves was enthusiastic. "I rejoice," the Wanderer says in *The Excursion*, "Measuring the force of those gigantic powers / That, by the thinking mind, have been compelled / To serve the will of feeble-bodied Man" (*Ex.* 8:204–07). From an aesthetic point of view, moreover, industrial landscape could be as sublime as London or the Thames. In 1812, Wordsworth described "the manufacturing Region" around Birmingham with vivid wonder:

> The whole space between W[olverhampton] and Birmingham is scattered over with towns villages and factories of one kind or another. Twenty years ago I passed over this same tract, and could then see in it nothing but disgusting objects; now it appeared to me (considered merely as a spectacle for the eye) very grand and interesting. The immense quantity of building spread over every side suggested the idea of Rocky Country, or an endless City shattered and laid waste by conflagration. The afternoon Sun played nobly upon the huge columns, and on the bodies of smoke that everywhere magnified or half obscured the various objects of the scene. In some spots also from very lofty Pipes like those of glass houses, flames of lively colour licked the air, as restless as the tongues of Dogs, when they are spent with Heat and hard running.[1]

There is disturbance in this scene, but also grandeur. What troubled Wordsworth more deeply than smokestacks were the social and economic changes associated with industrialization. After celebrating the "gigantic powers" appropriated by industrial engines, *The Excursion* goes on to warn that the system of factory labor enfeebles the bodies of youthful workers, breaks families apart, and corrupts "The old domestic morals of the land"

(*Ex.* 8:236). Wordsworth's letters on the Kendal and Windermere railway similarly object not to railways themselves—he had after all sought to invest in railways around Birmingham—but to the flood of weekenders a railway would bring to the Lake District (*LY* 1: 299–300; *PrW* 3:347, 3:355). Like many observers at the time, Wordsworth hoped that the benefits of mechanical power could be disengaged from the social effects that had so far accompanied them. His response to industrialization therefore has to be understood as a compound of conflicting impulses: cautious enthusiasm for "new-born arts," combined with fear of accompanying social change.

Many critics have written about Wordsworth's resistance to the social changes associated with industrialization; readers can now readily connect that resistance to the central passions of his poetry.[2] Less has been said about the other half of the compound—Wordsworth's embrace of industrial technology itself. Few critics have suggested that this enthusiasm significantly shapes the poetry at all. Wordsworth's poems praising mechanical power are not obscure; the sonnet on "Steamboats, Viaducts, and Railways" at least is widely anthologized, if the lines on steamboats from "To Enterprise" are not (*PW* 2:283). But readers tend to skim over these passages, because we interpret them—if my own experience is any guide—as routine concessive gestures made by a writer whose heart lies elsewhere. In the twenty-first century, the social importance of machine power is such a foregone conclusion that lines extolling it feel underdetermined. They seem to be dictated by history, rather than a distinct Wordsworthian voice or even a specifiable social perspective.

The importance of mechanical production was not quite as obvious in 1814 when Wordsworth published *The Excursion*, his longest and most complex response to industrialization. Although many writers claimed that new techniques of manufacture fueled British prosperity, others denied that those changes made much of a difference. Some economists who sympathized with landed interests held fast to the physiocratic argument that all production is at bottom agricultural. In 1808, William Spence declared *Agriculture the Source of the Wealth of Britain*, and denied "that manufactures *create* wealth" at all. "Though furnished with axes, and hammers, and trowels—with looms, and manufactures in profusion, it would be in vain to collect the necessary labourers, if food were wanting. . . . There is no difficulty in converting 100 quarters of wheat, by the intervention of the labour of man, into a steam engine; but no labour can transmute a steam engine back again into 100 quarters of wheat."[3] In the *Edinburgh Review*, Thomas Malthus criticized Spence, and argued that manufacturing did contribute something to the nation's net revenue.[4] But even Malthus insisted that agriculture was more vital for national greatness than manufacturing. "We are so blinded by the showiness of commerce and manufactures, as to believe, that they are almost the sole cause of the wealth, power, and prosperity of England. But perhaps they may be more justly considered as the consequences than the cause of this wealth."[5] The landowning classes (landed gentry and small proprietors alike) had strong reasons to minimize the importance of manufacturing and to insist on the primacy of agricultural production. Arguments like those advanced by Spence and Malthus had been used to justify existing laws that protected British farmers (and rentiers) against imported grain. As *The Excursion* went to press, similar arguments were used, by Malthus among others, to justify a new bill that would erect an even higher wall of protection.[6] Since Wordsworth expressed considerable interest in the Corn Bill of 1815 and for the most part approved of it, it is surprising that *The Excursion* does not parallel those arguments for the primacy of agriculture, but on the contrary attributes British prosperity to innovations in commerce and manufacturing (*MY* 2:209–11, 2:219–20). The Wanderer even traces agricultural prosperity itself to the stimulus of industrial innovation: "How much the mild Directress of the plough / Owes to alliance with these new-born arts!" (*Ex.* 8:131–32).

Wordsworth's praise of industry in *The Excursion* is a concessive gesture, then, but not a predictable or routine one. He waxes more enthusiastic than necessary, and admits premises that other writers who shared his political outlook would never have admitted. His admiration of industrial mechanism seems uneasy, but genuine, and deserves closer attention than it has yet received.[7] The admiration in fact emerged from habits of thought and feeling that were central to Wordsworth's poetic project. When the Wanderer "exult[s] to see / An intellectual mastery exercised / O'er the blind elements; a purpose given, / A perseverance fed; almost a soul / Imparted—to brute matter," he is defining mechanical production in terms that echo Wordsworth's own definition of poetic labor (*Ex.* 8:200–207). Like the engines that appropriate Nature's "gigantic powers," the poet aspires to possess "A power like one of Nature's" (*1805* 12:312). And like those engines, the poet humanizes nature's power by infusing it with "purpose" and "perseverance." Recollection in tranquility is necessary in part because it engraves that sense of purpose on the world: "habits of meditation have, I trust, so prompted and regulated my feelings, that my descriptions of such objects as strongly excite those feelings, will be found to carry along with them a *purpose*" (*PrW* 1:127). These parallels hint that industry, no less than poetry, carries on an "ennobling interchange" of power and purpose that shows how mind and world are fitted to each other (*PW* 5:5).

Wordsworth interpreted mechanical production, then, along lines shaped by his theory of imagination: he understood it as an appropriation and humanization of natural power. The analogy is not in itself very surprising. A writer who thought so regularly about poetry might have described almost any subject by extending a favorite metaphor from poetic theory. But in this case it is possible to show that the lines of influence flow in both directions. Wordsworth's accounts of industrial and poetic production parallel each other because they respond in parallel ways to an underlying social tension that had at least as much to do with industry as it did with poetry.

This tension was fundamentally a conflict between the landed gentry and the middle classes, but it manifests itself in Wordsworth's poetry as a recurring sense of economic insecurity and physical displacement. Describing his residence at Cambridge, for instance, Wordsworth acknowledges "fears / About my future worldly maintenance, / And, more than all, a strangeness in my mind, / A feeling that I was not for that hour, / Nor for that place" (*1805* 3:77–81). With characteristic precision, these lines fuse physical restlessness with social and economic unease. One might call this a consciousness of homelessness, with the proviso that Wordsworth was not literally threatened by homelessness in the twenty-first-century sense of the word: penniless exposure to the elements. What Wordsworth lacked (at Cambridge and for many years thereafter) was property, security, and social consequence.[8] I use the term "homelessness" to describe this feeling because it captures Wordsworth's insistence on interpreting those absences physically—as the absence of a place properly his own. That interpretation may have been shaped partly by biographical accident: the early death of both parents forced William to shuttle between two sets of relatives, and could certainly have fostered a desire for physical stability.[9] But Wordsworth's interpretation of social insecurity as displacement was also shaped—as David Simpson has elegantly shown—by his attachment to an agrarian model of community that represented civic responsibility as ownership and physical occupation of land.[10] Even after settling in Grasmere, and after his long-deferred inheritance materialized in 1802, Wordsworth's sense of rootlessness faded slowly. It was on one level a longing for a more palpable sort of permanence than a literary life in rented houses could ever provide. To put this another way: Wordsworth's recurring sense of homelessness reveals his uneasy station on the boundary between the landed gentry and the professional middle class. His agrarian ideals and family connections to the gentry seem to have led him to feel that there was something insubstantial and insecure about his own middle-class existence.

In Wordsworth's poems the theme of homelessness is not mainly an occasion for melancholy or nostalgia. By embracing the peripatetic vocation of poetry, he transforms the threat of dispossession into a privilege. The poet has no place that is properly his own because he is at home everywhere; he wanders in order to "apprehend all passions and all moods / Which time, and place, and season do impress / Upon the visible universe, and work / Like changes there by force of my own mind" (*1805* 3:85–88). The place of property is supplied by imaginative power, the portable form of capital he uses to "work . . . changes" on the universe, wherever he goes. This vision of resourceful independence admittedly remains shadowed by loss. Although imagination is strictly speaking placeless and motionless, its ubiquity often takes on overtones of exile. In the climax of *The Prelude*, to take a familiar example, imagination is embodied in a "homeless voice of waters" rising through a particular chasm in the mist—an image that para-doxically places the imagination while defining it as displaced (*1805* 13:63). This image condenses divided feelings that are dramatized more explic-itly elsewhere—in "Michael," for instance, and in the first book of *The Excursion*—as a tension between agrarian settlement and solitary imaginative enterprise.

In the last two decades several critics have explored that tension, showing in particular how Wordsworth's anxieties about property shaped his ambivalent response to vagrants, pedlars, and gypsies, who came to rep-resent both his hopes and his fears about a literary career.[11] This chapter extends that discussion by pointing out that Wordsworth's representations of industry dramatize a similar ambivalence. Wordsworth understood industri-alization, to be sure, as a threat to rural community. But in that respect his account of industry resembles his account of poetry more closely than is often realized. I have already mentioned that Wordsworth's poetic vocation was shadowed by a fear of homelessness: "Far from the world I walk, and from all care / But there may come another day to me—/ Solitude, pain of heart, distress and poverty" (*PW* 2: 236). Imagination promises to replace this lost security with a more deeply founded (because more portable) kind of independence. In *The Excursion*, industry likewise disrupts community and repays the loss with mobility. Although mills themselves are of course stationary, the poem understands them as manifestations of a restless power that is mobile both in the sense that it belongs everywhere, and in the sense that it permits or enforces human mobility. The Wanderer in fact directly compares the independence of an imaginative mind to the permanence and ubiquity of the forces that power industrial engines. Wordsworth's account of industrialization is admittedly darker, in general, than his narrative of poetic development. But they are versions of the same story: the story of a fortunate fall that prevents a protagonist from relying on local attachment, forcing him to cultivate deeper, placeless resources.

That narrative is Wordsworth's way of addressing the theme this book has been tracing since chapter 2: an analogy between middle-class enterprise and the protean universality of natural force. Wordsworth articulates the analogy less triumphantly and more dialectically than the reformist poets who celebrated middle-class energy in the 1790s. His poems characterize middle-class enterprise by comparing it to natural force, but they also express anxiety about social dislocation. The difference of tone is related to the ambiguity of Wordsworth's class position. His father had been an agent for the chief landowner of the district, and in later life William's own fortunes and political sympathies remained closely tied to the local gentry. His celebration of middle-class independence was thus qualified and darkened by a belief in the moral significance of landed property.

The heightened complexity of this issue in Wordsworth's poetry also reflects growing public consciousness about its economic underpinnings. The tension between local settlement and mobile power in Wordsworth's poetry echoes a contemporaneous debate between rival camps of political economists. Economists who sympathized with landed interests represented the cultivated field as a unique site of cooperation between human work and the productive power of nature. Thomas Malthus used that premise, for instance, to imply that rent (unlike other forms of profit) was a payment for the power of nature itself—a premise that seemed to prove that rent was fixed by natural law.[12] Economists like J.-B. Say and David Ricardo, who felt more sympathy for manufacturers, argued that nature's productive powers were ubiquitous, contributing equally to human enterprise in all locations. Land appeared to have a special status only because it was immobile and subject to local variation—facts, Ricardo argued, that should actually be viewed as "imperfections."[13] In economic discourse, as in Wordsworth's poetry, conflict between the landed gentry and the middle classes thus manifested itself as a debate about the relative advantages of fixed property and mobile power. The debate began in the late eighteenth century, but in Britain it became especially charged in the years leading up to, and following, the Corn Bill of 1815; in Wordsworth's poetry, it reaches a climax in *The Excursion* (1814). This chapter explores a parallel between Wordsworth's, and Ricardo's, reasons for stressing the ubiquity of nature's productive power. By the second decade of the nineteenth century, Wordsworth's political allegiances admittedly lay with landowners. But his poetry continues to resonate with a broader range of social perspectives than his overtly political statements might suggest.

Capitalism as Productive Wandering

The eighth book of *The Excursion* may seem an unlikely place to begin looking for Wordsworth's affinity with the economic interests of the commercial middle classes, since it is best known for its image of a mill lit garishly for nighttime toil, "where is offered up / To Gain, the master-idol of the realm, / Perpetual sacrifice" (*PW* 5:271). This would seem to be a moment where the author sternly and unambiguously criticizes the "getting and spending" of English commerce. The well-known image of the mill receiving and "disgorging" laborers is admittedly nightmarish, and Wordsworth describes the blight of child labor compellingly. His indignation deserves respect. But it is not the only emotion in this book of the poem. The illuminated mill is embedded in a larger narrative that has the form of a progress poem, following English commerce from its pastoral beginnings to a state of metropolitan grandeur. The narrative occasionally suggests a fall from innocence, but it also has overtones of promise and responsibility that evoke the end of *Paradise Lost* ("The world was all before them, where to choose / Their place of rest").[14] One aspect of the narrative that heightens those overtones is its personification of commerce as a Wanderer.

The Excursion as a whole is a loosely structured dialogue between the narrator, his friend the Wanderer (a retired pedlar), and a skeptical Solitary. In the opening pages of Book Eight, the Solitary introduces the topic of commerce as part of a flattering comparison he draws between the peripatetic life of a Knight buried in a churchyard before them and the "itinerant profession" of the Wanderer. "Errant those, / Exiles and wanderers—and the like are these; / Who, with their burthen, traverse hill and dale, / Carrying relief for nature's simple wants" (*Ex.* 8:45–48). Though some might scorn the "gross aims" of pedlars, for the Solitary these men epitomize the civilizing power of trade:

> Versed in the characters of men; and bound,
> By ties of daily interest, to maintain
> Conciliatory manners and smooth speech;
> Such have been, and still are in their degree,
> Examples efficacious to refine
> Rude intercourse; apt agents to expel,
> By importation of unlooked-for arts,
> Barbarian torpor, and blind prejudice;
> Raising, through just gradation, savage life
> To rustic, and the rustic to urbane. (*Ex.* 8:62–71)

This is a sanguine, Humean vision of commerce: "ties of daily interest" soften manners and elevate the tone of rural society. The pursuit of commercial wealth, in this humble form, does not seem unnatural; indeed, nature is the most portable form of the pedlar's portable wealth.

> to these poor men
> Nature (I but repeat your favorite boast)
> Is bountiful—go wheresoe'er they may;
> Kind nature's various wealth is all their own. (*Ex.* 8:58–61)

Up to this point the poem's argument will be familiar. Many critics have noticed that Wordsworth naturalizes his excitement about capitalist enterprise by focusing it on the figure of the pedlar. "Peddling," as Alan Liu aptly remarks, "is the Wordsworthian capitalization."[15] But other kinds of capitalization excited him as well. After thanking the Solitary for his praise, the Wanderer responds that a change has come over England, and over commerce. The pedlar's "foot-path faintly marked" has been displaced by canals and by "stately roads / Easy and bold." Hamlets have expanded into huge towns, "hiding the face of earth for leagues" (*Ex.* 8:121). This new commerce is industrial, and the Wanderer's evident unease might suggest that the poem will reject it as the antithesis of the pedlar's natural capitalism. But in fact the accent falls on continuity. Since the Solitary has just praised pedlars as agents of a civilizing process that raises "savage life to rustic, / And the rustic to urbane," the changes the Wanderer describes appear to be logical extensions of the pedlar's vocation. The Wanderer seems himself to initiate the civilizing process that displaces pedlarism and drives him onward.

The poem confirms that continuity by depicting industry as the pedlar's own itinerant life writ large. The pedlar's reliance on "kind nature's various wealth" is echoed, for instance, by this description of canals:

> The Earth has lent
> Her waters, Air her breezes; and the sail
> Of traffic glides with ceaseless intercourse,
> Glistening along the low and woody dale;
> Or, in its progress, on the lofty side
> Of some bare hill, with wonder kenned from far. (*Ex.* 8:111–16)

Here "traffic" seems interwoven with nature. Objectively it relies on waters and breezes "lent" by the earth. Subjectively the gliding sail is integrated into the landscape—disturbing the stillness only enough to elicit surmise (expressed grammatically by the "Or" that enumerates scenic possibilities). Even "the smoke of unremitting fires," which might have darkened this picture, "Hangs permanent, and plentiful / As wreaths of vapour glittering in the morning sun" (*Ex.* 8:125–28).

The mood darkens only when the Wanderer turns to describe the mill illuminated all night for "never-resting Labour's eyes" (*Ex.* 8:168). Here indeed he sees an "outrage done to nature" (*Ex.* 8:153). But the nature outraged is primarily human. The youthful worker who sacrifices health, dignity, and delight to his "close tasks, and long captivity" is maimed, in the Wanderer's view, both physically and emotionally (*Ex.* 8:296). The Wanderer has by contrast no objection to the instrumental treatment of inanimate nature; he celebrates it.

> I exult,
> Casting reserve away, exult to see
> An intellectual mastery exercised
> O'er the blind elements; a purpose given,
> A perseverance fed; almost a soul
> Imparted—to brute matter. I rejoice,
> Measuring the force of those gigantic powers
> That, by the thinking mind, have been compelled
> To serve the will of feeble-bodied Man. (*Ex.* 8:199–207)

Readers who expect Wordsworth's concept of nature to have a profoundly anti-industrial orientation have found this passage so odd that they have read it as insincere or ironic.[16] But the passage is consistent with the main tendencies of Wordsworth's thought. The humanization of nature is one of the strongest impulses of Wordsworthian lyricism. As early as MS. D of "The Ruined Cottage," Wordsworth's speakers struggle to glimpse a "secret spirit of humanity" hidden in nature's "calm oblivious tendencies," and look forward to a time when "All things shall speak of man."[17]

To be sure, the humanization of nature was only one half of the process that interested Wordsworth; he felt it ought to be accompanied by a reciprocal naturalization of human life. "Nature," in other words, was a name both for the physical world and for a moral ideal. In the latter sense it evokes, among other things, the value of spontaneity and autonomy—the value of a life in which "the essential passions of the heart . . . can attain their maturity, are less under restraint, and speak a plainer and more emphatic language" (*PrW* 1:125). This is why the Wanderer feels reservations about English industrial preeminence. English factories fall short of the moral ideal that nature represents, because labor in the factories prevents workers from developing their full range of human capacities. This is particularly criminal in the case of workers who are children. Because the chains are "Fixed in [the] soul, so early and so deep" the youthful worker "is a slave to whom release comes not, / And cannot come" (*Ex.* 8:299–302). But as the Solitary hastens to point out, much the same thing could be said about agricultural toil, with its accompanying poverty and ignorance. The Wanderer does not refute the remark. Wordsworth's empathy with a working-class point of view always had limits, and in *The Excursion* he reaches those limits more rapidly than in some of his earlier poems. He can sympathize with laborers, but he extends full respect only where he perceives "liberty of mind" (*Ex.* 8:433). In practice this means, to families that control at least a portion of the means of production—like Robert's loom, in Book One.

At this juncture a pair of boys enter the poem. One of them is the pastor's son, but the dramatic reason for their entrance is to reaffirm the possibility of work that cooperates with nature, and develops human capacities instead of deforming them. Interestingly, in describing this naturalized labor, Wordsworth draws on the same imagery he used to characterize the expansion of English industry and commerce earlier in Book Eight—an appropriation of natural power, which lends itself to human purposes. The boys are

> A few short hours of each returning day
> The thriving prisoners of their village-school:
> And thence let loose, to seek their pleasant homes
> Or range the grassy lawn in vacancy;
> To breathe and to be happy, run and shout
> Idle,—but no delay, no harm, no loss;
> For every genial power of heaven and earth,
> Through all the seasons of the changeful year,
> Obsequiously doth take upon herself
> To labour for them; bringing each in turn
> The tribute of enjoyment, knowledge, health,
> Beauty, or strength! (*Ex.* 9:259–70)

The passage balances between two of Wordsworth's characteristic passions. It celebrates a certain legitimate truancy, but remains at the same time committed to purposeful work, and subtly anxious about the possibility of slackening. Here, as in "Resolution and Independence" and many of Wordsworth's other poems, the purposeful restlessness of nature reconciles the competing claims of freedom and labor. The boys are "idle" without loss, because their wandering follows the apparent aimlessness of "the changeful year," and like the ceaseless change of the seasons is covertly a form of growth. In that sense the various powers of heaven and earth take it on themselves "to labour for them." There is an implicit contrast to the youthful inmates of the factory, who are enervated and weakened by their own diligence. But the metaphor of natural power as a substitute laborer echoes the poem's description of industrialization, and blocks any systematic nostalgia for a preindustrial world. One might attempt to preserve a nostalgic reading by arguing that nature volunteers itself as a substitute laborer for the schoolboys, whereas in the case of industry nature is "*compelled / To serve the will of feeble-bodied Man*" (my italics). But since the forces of nature are not sentient creatures, it is difficult to see the point of that distinction. In any case the poem does not consistently maintain it. In describing the industrial commerce carried on by canal, the Wanderer represented the powers of nature as volunteer workers ("The Earth has lent / Her waters, Air her breezes").

I believe that the parallels between Wordsworth's description of schoolboy idleness and his description of industrial mechanism are not ironic, but signal a real congruence of ideas. The industrial appropriation of natural force dovetails well with Wordsworth's tendency to imagine nature as a zone where errant passivity can become unconsciously active and provident. Wordsworth admired packmen and rural schoolboys because he saw in them gainful occupation without constraint; wherever they roam, "kind nature's various wealth is all their own." In a better-governed society, *The Excursion* implies, the industrial appropriation of water and steam would realize a similar aim. Diffusing human "purpose" and "perseverance" into the natural world, it would free human beings for a productive sort of vagrancy. Wordsworth praised joint-stock companies for just this reason in the 1835 "Postscript" to *Yarrow Revisited and Other Poems*. "[T]hey would enable a man to draw profit from his savings, by investing them in buildings or machinery for processes of manufacture with which he was habitually connected. His little capital would then be working for him while he was at rest or asleep. . . . " (PrW 3:248). For the employee, capital invested in his own company plays the role that the genial powers of nature play for the schoolboy: a substitute laborer, it permits rest without delay, harm or loss. The contrast between the captives of the factory and the roaming schoolboys is thus not really a contrast between an industrial and a preindustrial world; it is a contrast between the economic foundations of working-class and of middle-class life. The "gigantic powers" of nature that serve in the factory are also the "genial powers" of capital that ensure middle-class independence and permit productive vagrancy.

The Political Significance of Nature's Labor

Wordsworth's invocation of industrial capital as externalized, unconscious, tirelessly productive labor power may not surprise twenty-first-century readers. But in the early nineteenth century, this metaphor was charged and controversial. According to the physiocratic economists of the preceding century, land was the only genuinely productive form of capital. Only in agriculture could labor produce more than it consumed; all other classes were "unproductive," and subsisted on the surplus created by cultivators of the soil. The only way to externalize productivity was thus to invest in improvement of land. Adam Smith contested part of this physiocratic premise, claiming that the labor of merchants and artificers was after all productive. But even he conceded that agricultural labor was more so.

> No equal capital puts into motion a greater quantity of productive labour than that of the farmer. Not only his labouring servants, but his labouring cattle, are productive labourers. In agriculture too, nature labours along with man; and though her labour costs no expense, its produce has its value, as well

as that of the most expensive workmen. . . . [R]ent may be considered as the produce of those powers of nature, the use of which the landlord lends to the farmer. . . . It is the work of nature which remains after deducting or compensating everything which can be regarded as the work of man. . . . No equal quantity of productive labour employed in manufactures can ever occasion so great a reproduction. In them nature does nothing; man does all; and the reproduction must always be in proportion to the strength of the agents that occasion it.[18]

Labor, in Smith's view, is the source of value. So the labor of nature (added over and above the labor of man) necessarily makes agriculture more productive than other enterprises. This passage made "the work of nature" a central topic of economic debate for several decades. The concept is clearly ideological. One does not have to be a Marxist to see that inanimate nature does not work in the same sense that human beings do; discourse about the work of nature was in practice a way of justifying (and idealizing) the income that accrued from particular kinds of private property. Smith certainly felt that the owners of all forms of capital were entitled to a profit: "as something can every where be made by the use of money, something ought everywhere to be paid for the use of it."[19] But in his view, only landowners were entitled to say that they were paid for the labor of nature itself. Smith was not a protectionist, but writers arguing for tariffs on imported corn appropriated this passage to show that domestic agriculture (as the most productive human enterprise) deserved special attention and protection.[20] Abstract arguments about nature's role in production thus acquired an immediate political significance.

Writers who wanted to ascribe an equal productivity to industrial capital needed to counter Smith's argument that in manufactures "nature does nothing; man does all." They accordingly found ways of showing that nature labors not only in the field but in the factory. Alexander Hamiton's "Report on Manufactures" (submitted to the U.S. Congress in 1791) called for protective duties to foster manufacturing in the new republic. In the report, Hamilton devotes several pages to a refutation of Smith's argument about the superiority of agriculture, arguing in particular that the laws of mechanics constitute an "Agency of nature . . . made auxiliary to the prosecution of manufactures."[21] The French economist Jean-Baptiste Say, another advocate of industrial development, pursued a similar strategy. His *Traité d'Économie Politique* (1803) includes a chapter "Of the work of man and the work of nature," which defines the work of nature broadly enough to prove that it contributes equally to all human enterprises.

When I say *nature*, I mean to include all the material beings that compose the world. Each has its properties; all, or almost all, are able to contribute to productions useful to man. It is thus that fire softens metals, that wind turns our windmills, that water, air, and earth form the plants and trees that are useful to us. The elasticity of steel allows us to form springs that operate clocks; the weight of bodies serves us to the same purpose; we turn to our profit all the laws of the physical world. We are almost always working in concurrence with nature.[22]

But the arguments of Hamilton and Say by no means ended this debate. In England, Thomas Malthus was still quoting Smith in 1815 in order to affirm the superior productivity of agriculture and the necessity of a new Corn Bill.[23] In a pamphlet opposing the Corn Bill, and later in *Principles of Political Economy* (1817), David Ricardo responded by criticizing Smith's argument. "Does nature nothing for man in manufactures? Are the powers of wind and water, which move our machinery, and assist navigation, nothing? The pressure of the atmosphere and the elasticity of steam, which enable us to work the most stupendous engines—are they not the gifts of nature?"[24] Hamilton, Say, and Ricardo were among the first writers to systematically theorize industry as an appropriation of nature's productive power. That definition of industry went on to have a complex and varied nineteenth-century life; it serves different social interests in Thomas Carlyle, for instance, than it does in Friedrich Engels. But in debates about trade from 1791 through 1820, the political significance of the idea was clear and consistent. Liberal economists emphasized the industrial appropriation of natural force in order to undermine landowners' idealization of agriculture, which often hinged on its connection to the abstraction "nature."

In the spring of 1815, when Wordsworth became interested in debates about the importation of grain, his opinions were moderately conservative. He supported the spirit of the 1815 Corn Bill (which barred importation until the price of grain rose to 80s. a quarter), although he worried that 80s. might be a slightly excessive price (*MY* 2:219). The bulk of Book Eight of *The Excursion* was written earlier, between 1809 and 1812; it does not directly comment on agricultural protection, which was largely a moot question during those years of war. But Book Eight does express skepticism about economists who see industrial employment as a fair exchange for rural freedom: "Economists will tell you that the State / Thrives by the forfeiture—unfeeling thought, / And false as monstrous!" (*PW* 5:274).

So it is surprising to observe that when Book Eight describes industry, it does so in terms of the very metaphor liberal economists were using to prove industry's equality with agriculture. Industrial machines are not merely (as writers like Spence had argued) useful ways of abridging labor and "working up" the material provided by agriculture—they are for Wordsworth an expression of nature's primary productive power. Wordsworth may not have felt that this interpretation of machinery was politically significant. Economists had, after all, no monopoly on the nascent metaphor that made industry a "harness" thrown over the forces of nature, and Wordsworth is more likely to have encountered it in the newspapers—or in James Boswell's account of his visit to Matthew Boulton's factory—than in the works of Hamilton or Say.[25] But wherever he encountered the metaphor, it is significant that it caught his attention. It did so, I think, because the fantasy of harnessing natural force was closely related to Wordsworth's poetic project. Wordsworth's response to intimations of homelessness was often to imagine a way of finding security in displacement itself. In *The Prelude* this security takes the form of imagination. *The Excursion* imagines that pedlars find the same security in their itinerant life, which makes them proprietors of "Kind nature's various wealth." To interpret industry as a harnessing of natural force was to envision a similar source of security: a productive power that is invulnerable to dispossession, because it is located everywhere.

There is obviously an element of fantasy involved in this interpretation. Natural force may be placeless, but industry itself is not. The water-powered mills described in *The Excursion* were in reality confined to the banks of streams. Wordsworth was quite aware of their hunger for water power, and the threat it implied to his own mountainous and well-watered district; looking back in 1843, he was grateful that "the agency of steam" had diverted industrial development "to open and flat countries abounding in coal" (*PW* 5:468n). But social fantasy is not always closely tethered to fact. In the twenty-first century, personal wireless communication requires in reality a complex infrastructure, supported by massive investment. That fact has not prevented advertisers from representing wireless technology as an ethereal, mobile, and decentralizing principle. Early-nineteenth-century descriptions of industry are similarly fascinated by the image of a placeless power that permits human beings to brave the vagaries of location and climate. To quote Wordsworth's own lines on steam navigation: "adverse tides and currents headed / And breathless calms no longer dreaded, / In never-slackening voyage go / Straight as an arrow from the bow" (*PW* 2:283). Though mills rather than steamboats were at issue, Wordsworth's invocation of industrial mechanism in *The Excursion* is colored by the same fantasy.

The Luciferic Excursions of Imagination and Industry

Consider the role industrial imagery plays in Book Four, "Despondency Corrected." The title describes the Wanderer's effort to lift the Solitary out of a state of dejection brought on by the death of his wife and children and by disappointment in the French Revolution. The Solitary becomes a skeptic, whose vice is despair; he has resigned "All act of inquisition whence we rise / And what, when breath has ceased, we may become" (*Ex.* 3:235–36). His highest aspiration is "security from shock of accident" (*Ex.* 3:363). As Annabel Patterson has observed, the Solitary presents himself as "the happy Stoic recluse of georgic tradition." The poem's disdain for his Stoic retirement is a symptom of its broader dissatisfaction with georgic solutions; *The Excursion* is not a poem that celebrates or achieves contentment.[26] To relieve his friend's apathy, the Wanderer prescribes a life of risk-taking for which the general name is "imagination." The goal is not to bring the Solitary back to a straight and narrow path, but to encourage him to stray more widely—to exercise more of what the Wanderer calls, in a metafictional pun on the poem's title, "the mind's *excursive* power" (*Ex.* 4:1263).

The imaginative excursions the Wanderer prescribes take the form of physical excursions over the local hills. The Solitary came originally, like the Wanderer, from humble origins in Scotland. In the north of England they are both displaced figures. But the Wanderer stresses that English hills nevertheless provide a reservoir of energy that can be tapped to reawaken lost aspirations.

> Compatriot, Friend, remote are Garry's hills,
> The streams far distant of your native glen;
> Yet is their form and image here expressed
> With brotherly resemblance. Turn your steps
> Wherever fancy leads; by day, by night,
> Are various engines working, not the same
> As those by which your soul in youth was moved,
> But by the great Artificer endowed
> With no inferior power. (*Ex.* 4:550–58)

The soul-moving power of hills and streams is compared here to mechanical power. The point of the comparison is to show that the Solitary, though far from Scotland and without immediate family, possesses even in his exile an indestructible patrimony. Imagination need never be idle, because the engines that power it are at work constantly and everywhere. The metaphor seems concretely informed by Wordsworth's consciousness of industrialization. In 1814, "hills" and "streams" were in fact sources of industrial power; this is one reason why it makes figurative sense to describe them as psychological "engines." The phrase "by day, by night" suggests that imagination is not confined to any special time or season. But it also works more literally as part of the industrial vehicle of the metaphor. Water-powered engines could, after all, work uninterruptedly—by day or by night—with unfortunate consequences for the workers who tended them.

The Wanderer's vision of imaginative independence, in short, is built on a reassuring analogy to the ubiquity of industrial power. But the picture briefly evoked in these lines—of a Lake District where "every torrent and river" has its factory, lit all night for "never-resting Labour's eyes"—would appear to be a Wordsworthian nightmare (*PW* 468n, 270). Book Eight will make the author's opposition to night-work very clear. How is it possible that Book Four can invoke engines working night and day as emblems of the portability and permanence of imagination? I think the two passages are not as inconsistent as they might seem. A clue can be found in the way the Wanderer sets the scene for the description of the illuminated mill in Book Eight.

> When soothing darkness spreads
> O'er hill and vale, "the Wanderer thus expressed
> His recollections," and the punctual stars,
> While all things else are gathering to their homes,
> Advance, and in the firmament of heaven
> Glitter—but undisturbing, undisturbed;
> As if their silent company were charged
> With peaceful admonitions for the heart
> Of all-beholding Man, earth's thoughtful lord;
> Then, in full many a region, once like this
> The assured domain of calm simplicity
> And pensive quiet, an unnatural light
> Prepared for never-resting Labour's eyes
> Breaks from a many-windowed fabric huge . . . (*Ex.* 8:156–69)

The passage apparently contrasts the peace of evening against the unnatural light of the mill. But that contrast is complicated, in a strangely moving digression, by the emergence of the "punctual stars," who are part of the evening but (like the laborers) resist its homeward pull. "While all things else are gathering to their homes," the stars "advance," and "glitter," and remind "all-beholding Man" that he is "earth's thoughtful lord." The Wanderer reestablishes the ostensible moral point of the image by stressing that starlight is "undisturbed"; it is in that way distinguished from the mill-stream that "glares, like a troubled spirit, in its bed / Among the rocks below" (*Ex.* 8:179–80). But the reader's attention has already been caught, I think, by the stars' silent motion, and indifference to the call of home. Those images give nocturnal restlessness a dignity that carries over into the description of the "many-windowed fabric huge," making it less grotesque than darkly Miltonic—an effect underlined by Miltonic inversion of the adjective.

This is a moment that in fact closely fits David Simpson's thesis about the "Luciferic" overtones of *The Excursion*. To simplify his complex discussion: the ceaseless motion of the poem's central characters often seems shadowed by a consciousness of expulsion and exile. That shadowing consciousness is generated by the gap between Wordsworth's economic experience and his social ideals; the community of independent freeholders he failed to find in Westmoreland is projected into some other place or time, and imagined as loss.[27] But instead of settling into plaintive nostalgia, the poem typically embraces the grandeur of exile, struggling (as the narrator puts it in a moment of explicitly Miltonic reflection)

> To lift the creature toward that eminence
> On which, now fallen, erewhile in majesty
> He stood; or if not so, whose top serene
> At least he feels 'tis given him to descry. . . . (*Ex.* 5:298–301)

This language is less nostalgic than aspirational: the soul's memories of antenatal majesty might even (the narrator admits) be imaginative glimpses of an ascent yet to be tried. In my view this restless aspiration is the emotional center of the poem. When *The Excursion*'s emblems of lost community are examined closely, they often turn out to be things like pedlars, or wandering stars, that already carry within themselves a seed of exile. I do not mean to deny that Wordsworth's ideals about rural community were sincerely held. But where the poetry of *The Excursion* is concerned, alienation from that community seems to be a fortunate fall in a sense somewhat stronger than Milton would have admitted. The attraction of community hinges on the fact that community is already displaced.

The "punctual stars" that introduce the mill in Book Eight suggest that this principle holds true even for the explicit challenge to rural life posed by industrialization. The stars' "silent company . . . charged with peaceful admonitions" becomes a picture of the "calm simplicity / And pensive quiet" of an undisturbed rural community. But like the laborers streaming to the mill for the night shift, or the windows of the mill gleaming through the dark, the stars come out "while all things else are gathering to their homes," and the strange epithet "punctual" translates their endless voyage into a peripatetic vocation. The image of community, in short, is already mirroring the image of dislocation; and without that suggestion of something errant and off-balance, the image would not be (in this poem) quite as reassuring. The larger moral task of the narrative, after all, is to shake the Solitary out of his georgic lethargy, and to tempt him once again into a life of excursion. Wordsworth did not hope to see cotton mills multiply in Westmoreland; he grasped the human cost of factory labor. But it is not out of character for the Wanderer to make industry a consoling image of mobility in Book Four. Like other forms of homelessness, dark Satanic mills came to symbolize for Wordsworth a security deeper than the security of possession, founded on powers that were the more reliable for being restless. The "rumbling stream / That turns the multitude of dizzy wheels" may be a "troubled spirit," but the voice that rises from it has a solemnity akin to "a homeless voice of waters" (*1805* 13:63).

This structure of feeling was not originally a response to industrialization, but more broadly, an assertion of the social importance of the middle ranks of life. As chapter 2 has shown, middle-class radicals of the 1790s described their ambitions as natural energies in order to underline a contrast between the intriguing interest that governed a corrupt aristocratic system and the spontaneous claims of true merit. It was an oddly impersonal metaphor for personal aspiration, with a chilliness that comes through fully in a poem like Humphry Davy's "Life of the Spinosist," where ambition flows through the world, changing form ceaselessly, with "No sameness and no deep identity."[28] For Wordsworth, who felt drawn to an older agrarian model of community, that chill manifested itself more pointedly as a sense of dispossession—but a dispossession that is nevertheless embraced as independence and imaginative power. In *The Excursion*, this broad structure of class feeling is beginning to acquire a specific connection to industry. Although the poem criticizes the social consequences of industrialization, it also gazes admiringly at industry's connection to an infinite power that knows neither place nor season.

Wordsworth was not the only writer to think about industry in these terms. The Wanderer's emphasis on the ubiquity of natural force returns more polemically in David Ricardo's *Principles of Political Economy and Taxation* (1817). One of Ricardo's central purposes in that volume is to show that the foundational role commonly ascribed to agriculture does not belong to it, and belongs more properly (in one sense at least) to manufactures. From Adam Smith to William Spence and Thomas Malthus, economists had interpreted the fact that farmland returns a rent as an index of agriculture's surplus productivity. Rent, on this interpretation, measures the work contributed by the land itself, over and above the cost of labor and the ordinary profits on agricultural stock. Manufactures return no rent because "in them nature does nothing; man does all."[29] Rent thus proved that agriculture (and not incidentally, the class interests of the gentry) were securely founded on the indestructible productive powers of nature.

In the *Principles of Political Economy*, Ricardo reinterprets rent to prove that it is founded not on natural surplus but on limitation and exclusion. If land were "unlimited in quantity, and uniform in quality, no charge could be made for its use," for the same reason that no charge is made for the pressure of the atmosphere.

[W]ith the assistance of the pressure of the atmosphere, and the agency of steam, engines may perform work, and abridge human labor to a very great extent, but no charge is made for the use of these natural aids, because they are inexhaustible, and at every man's disposal. In the same manner the brewer, the distiller, and the dyer, make incessant use of the air and water for the production of their commodities; but as the supply is boundless, it bears no price.[30]

The supply of land, by contrast, is limited—and, even more crucially, parcels of land differ in their fertility and closeness to market. In Ricardo's analysis, it is this difference of productivity (rather than the power of fertility itself) that causes productive land to return a rent. As demand for produce increases, and poorer and more remote lands are brought into cultivation, the returns produced by a given investment of capital and labor on a given amount of land necessarily decline. It then becomes possible for the owners of fertile and conveniently located land to charge a rent equal to the difference between the returns for a given investment of capital and labor on their land and the returns for the same investment on the least productive land currently in cultivation. Ricardo uses this analysis of rent to overturn the traditional estimate of nature's role in agriculture.

[W]hen land is most abundant, when most productive, and most fertile, it yields no rent; and it is only when its powers decay, and less is yielded in return for labour, that a share of the original produce of the more fertile portions is set apart for rent. It is singular that this quality in the land, which should have been noticed as an imperfection, compared with the natural agents by which manufacturers are assisted, should have been pointed out as constituting its peculiar pre-eminence.[31]

In this manner Ricardo turns Smith's claim that nature works only in agriculture on its head. It is only because nature's assistance to agriculture is locally uneven and subject to decay that it is noticed at all. In manufactures, by contrast, nature's aid is "inexhaustible, and at every man's disposal." (The democratic implications of that phrase of course only extend to the middle classes—"given sufficient capital and know-how" is the unstated qualification.)

The Wanderer imagines nature similarly as a world of "engines" that are ubiquitous and at every man's disposal—given sufficient inner resources. Wordsworth's opinions about industrialization were quite different from Ricardo's; a figurative invocation of industry is not the same thing as an argument on its behalf. Still, both writers underline the placelessness of the forces at work in manufacturing, and both acknowledge that placelessness as one of the attractive features of the industrial economy. This habit of feeling will find more vociferous expression in the 1830s—for instance in *Sartor Resartus*, when Herr Teufelsdröckh exclaims that the "smithy-fire" of industry, "kindled at the sun," and "fed by air that circulates from before Noah's Deluge, from beyond the Dogstar . . . is a little ganglion, or nervous centre, in the great vital system of Immensity."[32] It mattered that the fire of industry should be "kindled at the sun" because early-nineteenth-century writers were describing industry in the context of a broader middle-class effort to validate mobile mercantile (and cultural) capital against the claims of landed property. Economists and poets emphasized industrial engines' connection to ubiquitous force because that was the perspective from which industrialization seemed most analogous to this broader struggle. Even a poet like Wordsworth, whose conscious opinions about industrialization were ambivalent, celebrated an analogy between the ubiquity of industrial power and the mobility of an imaginative mind.

Nature Writing as a Meditation on Industrial Power

Wordsworth's poetic relationship to nature thus turns out to be congruent, on several levels, with the relationship to nature imagined by early-nineteenth-century advocates of industrialization. Economists like Say and Ricardo insisted on manufacturers' connection to "the work of nature" because their audience took that connection as evidence of primary productivity and economic independence. Wordsworth insists on the same connection. The imaginative mind gratefully acknowledges the power it borrows from nature, and seeks to humanize that power by infusing it with "purpose" and "perseverance." A connection to forces that are at work everywhere reveals security and independence in occupations that might otherwise be interpreted as unproductive. Passive observers are not idle, and wanderers are not homeless, when "every genial power of heaven and earth / Through all the seasons of the changeful year, / Obsequiously doth take upon herself to labor for them" (*Ex.* 9:265–68).

I have traced this analogy in *The Excursion* because the industrial concerns of that poem make it possible to show that the parallel between Wordsworthian naturalism and middle-class economic discourse had substantial consequences: that it fostered a more sympathetic representation of industry, for instance, than *The Excursion*'s social argument might otherwise have dictated. But the class feelings that are revealed in *The Excursion* also shaped the representation of nature in earlier poems. Wordsworth's anxieties about property and place were, if anything, more acute between 1800 and 1802, just before the dispute over his inheritance was resolved.[33] The hope of finding security in homelessness itself—which always formed part of

Wordsworth's identification with nature—was in these years especially central to it. Even in the most playful works of this period—the poems to the daisy and lesser celandine, for instance—empathy for the living world is shot through with apprehensions of prodigality. Wordsworth addresses the lesser celandine, a humble roadside flower of early spring, in these terms:

> Ere a leaf is on a bush,
> In the time before the thrush
> Has a thought about her nest,
> Thou wilt come with half a call,
> Spreading out thy glossy breast
> Like a careless Prodigal;
> Telling tales about the sun,
> When we've little warmth, or none. (*PW* 2:143)

The observations are precise and moving because the speaker feels the element of risk involved in spring. To say that the thrush has no "thought about her nest" is to say that it's too early to venture much upon this season. With no more shelter than the thrush, the celandine nevertheless ventures everything—like a "Prodigal" (or a poet), "Telling tales about the sun / When we've little warmth, or none." Its ubiquity in "moor" and "wood" and "lane"—in any place, "Howsoever mean it be"—lifts the speaker's heart both because it proves that prodigality can be provident, and because in noticing the resourcefulness of this roadside flower he notices the portable resources of his own mind, which draw cheer from "the meanest flower that blows" (*PW* 4:285). Nature has a consoling power in Wordsworth's poetry largely because he identified in this way with the tenacious ubiquity of life, imagining that it could symbolically underwrite his own prodigal career. This economic dimension of Wordsworth's affection for nature becomes obvious in poems like "Resolution and Independence," where the character of the leech-gatherer is made to embody nature's tenacity and human independence at once. But even poems that seem relatively asocial—like those addressed to the daisy and small celandine—express similar anxieties and hopes (*PW* 2:135–39, 2:142–46).

A projection of an idealized middle-class independence onto fauna and flora in fact remained central to nature writing throughout the nineteenth century. John Muir pauses every few pages, in *My First Summer in the Sierra*, to admire the mobile self-sufficiency of the animals he observes. Since the grouse are "able to live on pine or fir buds, they are forever independent in the matter of food, which troubles so many of us and controls our movements. Gladly, if I could, I would live forever on pine buds, however full of turpentine and pitch, for the sake of this grand independence."[34] One can understand Henry David Thoreau's experiment at Walden Pond as a less literal version of the same project: an effort to found independence not on "a barn seventy-five feet by forty . . . and one hundred acres of land" but on a relationship to the inexhaustible productive force of Nature itself, glimpsed (for instance) in the leaflike shapes of sand and clay that ooze from a thawing bank in the climactic chapter "Spring."[35]

Advocates of industrialization like Say, Ricardo, and Carlyle were fired by visions of a similar relationship to nature's productive energy—a power that was, in Ricardo's words, "inexhaustible, and at every man's disposal." In Wordsworth's *Excursion* these two visions not only parallel each other but substantially overlap; the indifference to location Wordsworth admired in the celandine is expressed as the omnipresence of latent industrial power. I do not want to make too much of the irony involved in this convergence of nature writing and industrialism. It is primarily a retrospective irony, visible to twenty-first-century readers because a tension between industrial and environmental goals has given rise to some of the central conflicts of the present age. In the early nineteenth century, as Leo Marx pointed out in *The Machine in the Garden*, a writer like Emerson saw nothing unseemly about hitching his transcendental and pastoral ideals to the engine of industrial progress. Of course, Marx explains this as a uniquely American conjunction of opposites, made possible by Emerson's belief that technology would open up the American wilderness and transform it into a "middle landscape" of rural independence.[36] Marx used Wordsworth to exemplify an alternate, pessimistic, English response to industry—a response that already (and in Marx's view, correctly) recognized millwheels and mountains as antagonistic principles.[37]

This chapter has questioned that reading of Wordsworth. Admittedly, on balance, Wordsworth was less enthusiastic about industrialization than Emerson. But *The Excursion* is a complex poem. Amid its worries about the social consequences of the factory system, it foreshadows Emerson's enthusiasm, by using the latent industrial potential of air and water to represent the independence of an imaginative mind that can draw power from the world wherever it travels. The convergence of transcendental and industrial rhetoric in nineteenth-century nature writing is not a function of American geography. It reveals rather that writers on both sides of the Atlantic found it easiest to interpret the new industrial economy in terms of established patterns of class feeling. For many middle-class writers, natural force already symbolized a mobile form of independence that did not require possession of land; industrial mills and engines simply literalized that metaphor.

The pattern I have traced in *The Excursion* was set much earlier in Wordsworth's career. In the early 1790s, English radicals were already establishing an analogy between the energies of nature and the work done in the middle ranks of life. Wordsworth's identification with the homeless independence of spring flowers, or pedlars, or industrial engines, is his way of drawing the same analogy, in order to show that there is a natural foundation for personal autonomy that does not depend on land. Because Wordsworth had strong ties to other classes (sympathy with small freeholders, material dependence on the landed gentry), this analogy never became as absolute or as abstract in his poetry as it did in the writings of, say, William Godwin. It took instead a paradoxically concrete form—appearing as a flower whose obscure and homely location makes it a figure for homelessness, or a mill that is at once a place of confinement and a mirror for the wandering stars.

CHAPTER 6

SUNLIGHT AND THE REIFICATION
OF CULTURE

"To Keats," Ian Jack comments, "Apollo was more than a figure of speech or a literary allusion: he was becoming something very close to the God of his adoration."[1] Keats could treat Apollo as a living god partly because Apollo was understood (through his connection to the sun) as a god of energy and ambition. He was thus a fitting patron for Keats's social, as well as literary, aspirations. But Apollo was also particularly the god of the lyre, and Keats's seriousness about Apollo went hand in hand with a tendency to think of poetry as a power with a real existence outside written texts—and for that matter, outside writers and readers of poems. Keats often defines poetry as pure power—agency without an agent—for instance, in "Sleep and Poetry": "A drainless shower / Of light is poesy; 'tis the supreme of power; / 'Tis might half slumbering on its own right arm."[2] William Hazlitt's lecture "On Poetry in General" reifies poetry in a similar way. For Hazlitt, "the poetical impression of any object is that uneasy, exquisite sense of beauty or power that cannot be contained within itself," which leads him to the oddly concrete conclusion that poetry exists "wherever there is a sense of beauty, of power, or harmony, as in the motion of a wave of the sea, [or] in the growth of a flower that 'spreads its sweet leaves to the air, and dedicates its beauty to the sun.' " Here "poetry" is less a description of a particular human activity than a name for a generalized power that encompasses life itself. "Poetry," according to Hazlitt, "puts a spirit of life and motion into the universe."[3]

This reification of poetry might be taken as the distinguishing characteristic of Romantic criticism. Samuel Taylor Coleridge's doctrine of the primary Imagination drives toward this goal, albeit on a theological rather than empirical plane. A separation between the power of poetry and the mere written poem is equally central to Percy Bysshe Shelley's "Defence of

Poetry," which implies, in fact, that Poetry ceases to be present at the moment the poem is written down. German Romantic criticism comes to similar conclusions. Written poems are comparatively worthless, according to Friedrich Schlegel, compared to "the unformed and unconscious poetry that stirs in the plant and shines in the light."[4]

All these critical dicta reify poetry, because they consider it neither as a human activity, nor as a genre of literary products—but as a power with an existence separate from the poem and the poet. Although I borrow the concept of reification from Marxist tradition, I use the word here to describe rather than condemn. (Marx's own use of the verb *vergegenständlichen*, for that matter, is not consistently condemnatory.)[5] A language that separates nouns from verbs inevitably abstracts some patterns from the world and treats them as grammatical subjects. In the case of terms like *centrifugal force* or *poetry*, it is easy to see that this involves a figure of speech: a gentle form of personification. But nouns are always figurative. Even in the apparently uncontroversial cases of persons and moveable objects, there are trade-offs involved, as Marx's discussion of commodity fetishism makes clear: to treat the commodity as a independent entity is to risk blurring the social relations involved in labor. But then to reify labor-power, as Georg Lukács warns, is to risk blurring the individual experience of the worker. And to reify the individuality of the worker is to blur the conflict of social forces.[6] The line between abstractions and real objects will always remain contested: the Romantic reification of poetry is not inherently more figurative than, say, contemporary critical reification of "discourse." But like all figures of speech, reification has consequences, and the more literally it is taken, the more interesting the consequences.

The claim that "poetry puts a spirit of life and motion into the universe" now sounds reasonable only when it is read as a psychological claim. The power of poetry can then be understood as a shaping power in the poet's mind; the claim that it puts life and motion into the universe can be understood as a metaphor for the process of perception. Twentieth-century rehabilitations of Romantic theory tended to emphasize this psychological dimension of the period, and twentieth-century critiques of Romanticism tended to indict writers for subjectivist assumptions. But in the early nineteenth century the line separating psychological from physical powers could not be drawn with great precision. To understand Romantic poetic theory in its own terms, readers must first set aside anachronistic confidence about the boundary between psychology and physics, and believe that Hazlitt means it when he says that "poetry puts a spirit of life and motion into the universe."

To believe this is not to say that Hazlitt was an idealist. The axis that unites writers as different as Hazlitt and Mary Robinson, Keats and P. B. Shelley, cannot have much to do with vexed questions of ontology. These writers wouldn't have agreed about the distinction (if any) between matter and mind, but they do agree that poetry is a power with an existence outside the poet, and that it is manifest (whether materially or immaterially) in all life and motion. The distinction involved here is rather the distinction that Jerome Christensen draws in *Lord Byron's Strength* between "Romantic strength" and "empirical force." Christensen borrows his definition of "strength" from Hannah Arendt: "Strength unequivocally designates something, an individual entity; it is the property inherent in an object or person and belongs to its character, which may prove itself in relation to other things or persons, but is essentially independent of them."[7] Christensen uses "power" and "force," by contrast, to name a kind of agency that is relational and fungible, an agency that can be abstracted from its possessor and enter into circulation.

The distinction is a clear and useful one, but it is not clear where Romantic theories of poetry fall on this axis. In *Lord Byron's Strength*, Christensen takes Byron's effort to reimagine his own nobility as the paradigm of an aristocratic concern with personal strength; Christensen implies that this personal and nontransferable model of agency revives in the Romantic period to oppose the objectification of social relations that typifies political economy. The argument, though based on Byron's case, extends beyond him: "Resistance to objectification integuments Romantic practice," as Christensen puts it.[8] The examples considered in this chapter will qualify that conclusion. Romantic writers commonly reify poetry as power, and they are particularly fascinated by the fluidity of power, which allows it to circulate freely, without a determinate possessor. This model of poetic production parallels the account of industrial production offered by economists like J.-B. Say and David Ricardo, who celebrated the ubiquity of industrial power. Romantic poets did criticize political economy, and they did so in part—as Christensen rightly argues—by insisting on personal strength and struggle. But in the Regency decade and immediately thereafter, personal strength becomes interesting in poetry largely because of the pathos of its dependence on impersonal force. The personalities that draw a reader's attention do so by admitting that personality itself is only "a fading coal which some invisible influence, like an inconstant wind, awakens to transitory brightness." The fire that burns within the poet "is not . . . a power to be exerted according to the determination of the will"; nor is it a specific inspiration. It is "the awful shadow of some unseen Power," borrowing "grace" and "mystery" from its diffusion and mutability (*SPr* 294, *SP* 529–30). Although these phrases are drawn from Shelley's "Defence of Poetry" and "Hymn to Intellectual Beauty," the dramatic situation they evoke can be traced also in Byron's *Manfred*, L.E.L.'s "Lines of Life," Hemans's "Properzia

Rossi," Bryan Procter Waller's "Girl of Provence," and many other late-Romantic poems whose protagonists are consumed by a fire they cannot securely possess.

These texts explore dilemmas that emerge, in part, from a blurring of the boundary between power and strength. The poetic speaker often reaches a climax of self-possession—as in Manfred's final address to the sun—just at the moment he or she confesses dependence on an impersonal force that shaped consciousness and will destroy it from within. Christensen's distinction between power and strength valuably illuminates poems like these, by identifying their central paradox. But it is difficult to interpret that paradox as expressing a general "resistance to objectification"; the tragic overtones of the poetry often depend on the inimical grandeur of reified power. To admit this is not to indict the poems. Poetry cannot resist reification generally and absolutely; to use language is inevitably to reify some aspect of the world. The choice between strength and power is a choice between reifying a personal attribute, or agency itself. I see no basis on which to generally recommend one of those choices over the other. It does seem possible, on the other hand, to describe the social interests that were advanced by the specific reification of creative power in British poetry between 1814 and 1832.

The Regency-era metaphors that make creative power a self-consuming fire or a temporary gift of "life and light" in some ways resemble metaphors that British writers used a decade or two earlier to naturalize middle-class ambition. In the 1790s, writers with reformist sympathies relied on similar imagery—Promethean fire, electric spark, animating light. More importantly, they used that imagery to explore the same unstable boundary between "strength" and "power" that would occupy poets during and after the Regency. By concretizing Liberty as an almost physiological fountain of energy and fervor, Mary Robinson's "Progress of Liberty" (1798) defines self-determination, paradoxically, as an effect of participation in abstract power. In Humphry Davy's "Life of the Spinosist" (1800), analogies between human aspiration and natural force celebrate the universality of a mind with "No sameness and no deep identity, / Linked to the whole, . . . / Impressible as is the moving sky, / And changeful as the surface of the seas."[9] In the 1790s, as I argued in chapter 3, the point of universalizing ambition in this manner was to claim that middle-class work contained within itself a disinterested autonomy like the autonomy gentlemen supposedly derived from inherited land.

In the second and third decades of the nineteenth century, analogies between poetry and natural force similarly universalize aspiration. But the tone and scope of the analogy change in ways that suggest a different social function. The tone darkens. When Shelley, or Byron, or Hemans compare the mind's energy to solar fire, their emphasis tends to fall on its transitory, flickering, and self-destructive character. At the same time the scope of the analogy between human agency and natural force contracts. Writers of the 1790s compared sunlight to energy and enterprise; in the Regency era solar imagery more commonly evokes spiritual or artistic aspiration. This involves a narrowing of social scope, but the nature of the contraction needs to be defined carefully. I do not mean to imply any slackening of political commitment. "Poetry" is after all for Shelley a power "connate with the origin of man," expressed not only in "metrical language," but also in "music," "dance," "architecture," and "statuary," as well as law and religion (*SPr* 277, 279). Moreover, Shelley wrote poems that describe Liberty as "lightning" and "sunlight," and that attribute to it many of the same powers the "Defence" attributes to Poetry (*SP* 622). But though Shelley's analogy between liberty and natural force is passionately political, it does lack a certain social specificity. Writers of the 1790s foregrounded the economic dimension of liberty: the point of comparing liberty to natural energy was largely to evoke the frustrated expectation that "extraordinary talents and virtue" would be rewarded by "preferment."[10] To quote William Godwin on the bandits led by Mr. Raymond,

> Energy is perhaps of all qualities the most valuable; and a just political system would possess the means of extracting from it thus circumstanced its beneficial qualities, instead of consigning it as now to indiscriminate destruction. We act like the chymist who should reject the finest ore, and employ none but what was sufficiently debased to fit it immediately for the vilest uses.[11]

The bitterness of this analogy between human ambition and natural force registers a collective economic disappointment. That bitterness does not make Godwin's writing more generous or more principled than Shelley's— but it may more candidly reflect the social basis of political struggle.

Writers who came of age in the first decades of the nineteenth century, by contrast, elevate ambition to the dignity of a natural force only when it can be interpreted as artistic aspiration. Byron and Shelley wrote this way in part because material want was not in fact a pressing concern. But the same reticence is perceptible in poets who had to (or hoped to) write for a living— in L.E.L., in Keats, and even in John Clare. All three writers were candid both about their outsider status and about their yearning for poetic fame. But the yearning for fame is kept separate from the tabulation of "red-lin'd accounts."[12] They do not—like Godwin or Robinson—invoke an "energy" that is shared by all ambitious minds, linking literary labor to the struggles of talented clerks and banditti. John Clare writes at length, very frankly, about physical labor. But he doesn't understand his labor in the field as an expression of the same "kindling warmth" that "glows" in his verse; he compares only his literary labors to sunlight. Clare's stance contrasts markedly with the stance of earlier plebeian poets, who often found it difficult to justify their literary work outside the context of manual labor that gave it a social location.[13] An analogous change of tone is perceptible in middle-class writers, who increasingly separate literary aspiration from ordinary social climbing. Mary Wollstonecraft took it for granted that all people with "abilities and virtues" would want to ascend "from the middle rank of life into notice," and called this "rising by the exertions which really improve a rational creature."[14] L.E.L. states more delicately that "song has touch'd my lips with fire, / And made my heart a shrine; / For what, although alloy'd, debased, / Is in itself divine."[15]

Many different factors undoubtedly converged to foster this change of tone, including political retrenchment after the French Revolution and a concomitant reaffirmation of class hierarchy. But whatever it may say about the contraction of political possibility in the early nineteenth century, writers' increasingly single-minded focus on literary fame also reveals a new confidence that cultural attainment is intrinsically more valuable than other forms of distinction. To be touched by the fire of song is in itself, for L.E.L., sufficient praise. A heart that responds to poetic fire will necessarily contain feelings that rise above flattery and outlast ridicule. Regency writers' increasing insistence on separating literary aspiration from vulgar ambition can also be understood, in other words, as a discovery of the autonomy of culture. Artistic culture confers on its possessor a distinction that is increasingly understood as incommensurable with other possessions and forms of honor.

Here I am touching on the social transformation Pierre Bourdieu has described as the emergence of an autonomous cultural field—a topic closely related to the reification of poetic power in the early nineteenth century. Bourdieu resists Marx's assumption that classes are objectively given entities that merely await self-consciousness. He argues that classes are constituted as much by cultural and symbolic capital as by their relation to the material means of production.[16] A theory of class that emphasizes symbolic underpinnings has the potential to be all too attractive to disciplines that are primarily concerned with symbolic artifacts, like my own discipline of literary studies. I will try not to lean too hard on the phrase "cultural capital," which is after all a condensed metaphor, and bears only limited weight. But Bourdieu's analysis produces valuable insights into the lived experience of class. Classes are economic entities, but they are also products of symbolic contestation. Even when that contestation mystifies underlying economic relations, it has real social consequences: the mystification itself becomes part of the experienced reality of class.

According to Bourdieu, the relative importance of different kinds of capital does not remain constant across history; nor, in fact, do the divisions between them. The broad domain of symbolic capital—"accumulated prestige, celebrity, consecration, or honour"—for many centuries included a subdomain of artistic and literary attainment without distinguishing that domain as "culture."[17] Cultural distinction can correlate so closely with social distinction that the two are difficult to separate; this tends to be the case in an age—for instance Augustan England—that ranks cultural products by reference to the same standards of urbanity and decorum that regulate the social realm. But it is also possible for the social and cultural fields to be organized along independent lines. Culture then becomes perceptible as a source of prestige separate from general social refinement. Indifference to the criteria of social prestige may even become itself a source of cultural distinction. Bourdieu describes the decreasing correlation between the axes of social and cultural distinction as an increase in "the autonomy of the cultural field."[18]

The writings of Raymond Williams suggest that the ideal of artistic culture separated from the broader domain of cultivation between the middle of the eighteenth and the middle of the nineteenth century.[19] Trevor Ross has discovered early symptoms of this separation in mid-eighteenth-century attacks on classicism in favor of "pure poetry." Poetic theorists like the Wartons encouraged writers to draw directly on "natural genius" instead of admitting dependence on existing symbolic capital by mastering classical models or imitating contemporary manners. Their elevation of natural genius was a way of insisting that literary culture constituted its own source of distinction.[20] The "Preface" to Lyrical Ballads can usefully be viewed as a culmination of the tradition Ross describes. Wordsworth's commitment to the separation of cultural distinction from other social standards is evident both in his deliberate choice of "low and rustic" subjects and in his contempt for those who "converse with us . . . gravely about a *taste* for poetry, as they express it, as if it were a thing as indifferent as a taste for Rope-dancing, or Frontiniac or Sherry" (*PrW* 1:139). Poetry was not to be reduced to one class signifier among others; it constituted a realm of distinction unto itself. Wordsworth's polemical stance in the "Preface" also reveals, however, that in the year 1800 he still felt outnumbered on this issue. The concept of culture was not fully separated from other forms of distinction—and the word was not common in its modern sense—until the middle of the nineteenth century.[21]

I believe that the gradual emergence of an autonomous cultural field helps to explain the changing significance of solar mythography in the early nineteenth century. In the 1790s, as I argued in chapter 3, poets invoked sun-gods to define the autonomy of middle-class work. For that reason they borrowed a metaphor from recent natural philosophy and imagined the sun, not primarily as the light of day, but as a fountain of abstract, mobile agency. That metaphor persisted into the early nineteenth century, but its class affiliations became increasingly diffuse. It was taken up by peers of the realm (Byron) and by farmhands (Clare). At the same time its social scope became, in another sense, narrower: instead of using solar agency to define the autonomy of work in general, poets increasingly used it to define the autonomy of culture. The remaining pages of this chapter explain in more detail how scientific ideas about the sun were invoked to reify culture as a natural force.

"The Fire Which We Endure"

The sun-worship practiced by Shelley and Byron can look, at first, like a quirk of personality. Shelley's friends called him "the Salamander" because of his fondness for sunbathing. Byron frequently drops hints that suggest a kind of personal understanding with the sun. "I am always most religious upon a sunshiny day—as if there was some association between an internal approach to greater light and purity—and the kindler of this dark lanthorn of our external existence."[22] But sun-worship was not a private quirk; it was a public emblem of shared political and literary allegiances. Marilyn Butler has persuasively described the circle centering on Shelley, Byron, Thomas Love Peacock, and Leigh Hunt as a "Cult of the South"—taking these writers' fascination with the Mediterranean as a pole around which to organize their interest in mythography, celebration of sexual pleasure, critique of Christianity and the Lake poets, and revival of classical (especially comic) forms.[23]

This form of classicism centers less on the Acropolis or the Forum than on the warmth and light of the Mediterranean sun, which comes to emblematize a rejection of the political, religious, and stylistic gloom of counter-Revolutionary Germany and England, in favor of Apollo, god of "life and light." The choice of the sun as the group's emblem was not determined mainly by the geographical accident that Greece and Rome happen to be located south of Britain. More important was the sun's centrality in late-Enlightenment science and in the religion of dynamic naturalism that accompanied it. By taking the sun as their emblem, this group of genteel writers wrapped themselves in the mantle of an earlier generation's radicalism, and appropriated for poetry and culture a naturalistic myth that middle-class intellectuals had used to give their aspirations the universality of natural law. This section of the chapter, on Byron and his imitators, will explain how the imagery of solar aspiration was refashioned to express a doomed, self-consuming energy. The subsequent two sections, on Shelley, will show how that transfiguring but transitory light could expand into a figure for culture in general.

Solar myth plays a starring role in Byron's dramas; a soliloquy delivered to a setting or a rising sun often sets the tone at the beginning of an act. Act II of *Sardanapalus* begins with Beleses, priest of Baal, apostrophizing the setting sun that is also his god. Act V of the same play begins with Myrrha's soliloquy to the rising sun, "that which keeps all earth from being as fragile / As I am in this form." Placing these soliloquies at the transitional moments of sunrise and sunset allows the sun's arc to underline or to ironically reverse the arc of the speaker's own life: a character who apostrophizes the (rising or setting) sun in the last act of a play by Byron will always be dead by the end of the play. As often in the Romantic era, the implied figure of speech is metaphor on one level, but metonymy on another: the fading life is like the fading light, but also literally part of it. Byron's fascination with this doubleness is particularly clear in Manfred's soliloquy as he watches his last sunset.

> Most glorious orb! that wert a worship, ere
> The mystery of thy making was revealed!
> Thou earliest minister of the Almighty,
> Which gladdened, on their mountain tops, the hearts
> Of the Chaldean shepherds, till they poured
> Themselves in orisons! Thou material God!
> And representative of the Unknown—
> Who chose thee for his shadow! Thou chief star!
> Center of many stars! which mak'st our earth
> Endurable, and temperest the hues
> And hearts of all who walk within thy rays!
> Sire of the seasons! Monarch of the climes,
> And those who dwell in them! for near or far,
> Our inborn spirits have a tint of thee
> Even as our outward aspects;—thou dost rise,
> And shine, and set in glory. Fare thee well!
> I ne'er shall see thee more. As my first glance
> Of love and wonder was for thee, then take
> My latest look: thou wilt not beam on one
> To whom the gifts of life and warmth have been
> Of a more fatal nature. He is gone:
> I follow. (*BW* 4:94)

The first part of this speech is complicated by Manfred's hesitation between skeptical and Christian interpretations of solar myth. He notes, in the tradition of Charles-François Dupuis, that sun-worship preceded Biblical revelation, but also suggests that the sun is a representative created by "the Almighty" or "the Unknown." In the larger context of the play, it's far from clear that this remote creator-figure can be read as a Judeo–Christian god; the religious content of the passage is probably best described as a mixture of Shelleyan skepticism and deism.

But the myth that concerns me in this chapter takes shape with the remark that the sun so gladdened "the hearts of the Chaldean shepherds" that "they poured / Themselves in orisons!" The physicality of the image is underlined in Byron's manuscript version of the line— "it poured / Itself in orisons"—which suggests that it is not the shepherds understood as persons but their heart, or even the sun itself, that pours forth. This begins to hint that the human heart is a vessel for containing sunlight, a suggestion further developed by the claim that "Our inborn spirits have a tint of thee / Even as our outward aspects." "Tint" evokes the eighteenth-century theory that race and culture are products of climate; a theory that continued to shape geographical thought—as Alan Bewell has shown—in the Romantic period.[24] But that is not the only theory about the sun the passage explores. Our inborn spirits are shaped by the sun not only because of climate but, as the passage goes on to reveal, because the sun is physically the source of "life and warmth," and visually the source of "love and wonder." As in Davy's "Life of the Spinosist," pleasure and physical life are kindled simultaneously by the sun. And, as in Davy, "my first glance / Of love and wonder" suggests an analogy between the sun and the maternal breast.

If the passage celebrates the sun's parental responsibility for the human mind, it also uses that responsibility as an occasion for a mysterious self-pity. What can Manfred mean by saying that "the gifts of life and warmth" have been, for him, gifts of a "fatal nature"? One can imagine a story within which that remark would make a non-banal kind of sense: Victor Frankenstein, for example, could intelligently complain that the spark of life had been, for him, uniquely fatal. *Manfred* is not that story. Manfred's bitter remark nevertheless makes sense, because this sort of lyric gesture—evoking an unstated story about the solar origin and fate of human aspiration—was by 1816 quite familiar.

Davy's "Life of the Spinosist" provides a useful summary of the implied narrative. Davy's protagonist draws life, motion, and joy from "kindling spirits" the sun pours over the earth. But Davy's poem already contains a shadow of the fatal frustration Byron will explore. His protagonist feels "the social flame" and gives "man / Ten thousand signs of burning energy" before glimpsing "the nothingness of human things" and retreating to solitude. As in many Gothic subplots, the social hypocrisies that check the protagonist's energy remain obscure. The conflict between social inertia and pent-up ambition hints at a class struggle. But "The Life of the Spinosist" translates that specific frustration into a general mortal fate, sending the protagonist's social flame back to its physical source: "Etherial fire to feed the solar rays, / Etherial dew to feed the earth in showers."[25]

It has long been clear that the brooding heroes and Gothic villains of the 1790s lent some of their pent-up energy to Byron's protagonists—whose powers are similarly blocked, and become similarly self-destructive. In Byron's writing, this frustration has less overt economic significance than it does in (say) *Caleb Williams*. But the disappointment retains a social dimension— visible most obviously in Byron's disgust with the politics of restoration after Waterloo. More broadly, as Jerome McGann has suggested, Byronic flashes "of weariness or scorn" reflect ambivalence about the historicity of social existence, which forces even poets to trim their sails to the prevailing winds of politics or fashion.[26] Byron, like Davy, often seeks to translate this social frustration into a simpler tragedy of mortality. He does so, moreover, in much the same way Davy did—by drawing on a shared myth about the instability of vital power borrowed from the sun.

Davy helped to disseminate that myth widely—not through his poetry so much as through his scientific work. In the system of Brunonian medicine that flourished in the Romantic period, human physiology was compared to a self-consuming fire: the stimuli that made the fire burn brightly also consumed its fuel, so that intensity of life was itself a cause of disease and death.[27] Researchers like Davy and Thomas Beddoes connected this medical system to Antoine Lavoisier's theory of respiration, which seemed to imply that oxygen (and thus, indirectly, sunlight) provided the fuel for the fire. The same hypothesis was reproduced by a number of American physicians.[28] When Byron imagines creative energy as a self-consuming flame borrowed from the sun, he is echoing this widely disseminated medical theory. Thomas Moore does the same thing in his "Song of a Hyperborean." Greek tradition stated that the island of Hyperborea was consecrated to Apollo; Moore works this dry fact up to a distinctly Romantic poignancy.

> To the Sun-god all our hearts and lyres
> By day, by night, belong;
> And the breath we draw from his living-fires,
> We give him back in song.[29]

Poetic life is represented here as a condensed version of normal physiology. All mortals have to give back the breath they draw from the sun's "living fires." To give it back "in song" is only to hasten and intensify that common fate.

Arimanes, the Zoroastrian power of darkness, makes an appearance in *Manfred*, but his foil Oromazde, the power of light, never does. He doesn't have to appear, because his role is effectively filled by the sun's visible orb; and he shouldn't appear, because the sublimity of Manfred's solitude would be compromised if the sun he celebrates and defies were to be personified any further. But in works that are less dramatic, Byron has more freedom to project his contemporary solar myth onto ancient names. One such moment occurs toward the end of *Childe Harold's Pilgrimage*, in a passage that is significant both for the work as a whole and for the influence it exerted on other Romantic writers. It forms part of the climax of Canto IV, which attempts to represent art as a power that re-creates and redeems the ruinous processes of history and human mortality alike.[30] The narrator has been surveying Italian monuments (St. Peter's and the Vatican); in stanza 161, he turns finally to the Apollo Belvedere:

> Or view the Lord of the unerring bow,
> The God of life, and poesy, and light—
> The Sun in human limbs arrayed, and brow
> All radiant from his triumph in the fight . . . (*BW* 2:178)

Apollo was not in Greek religion the god of life, but rather the sender of plague and thus also the protector of health and patron of physicians.[31] Romantic writers' frequent characterization of Apollo as a god of "life and light" is a recent invention, shaped by assonance, but above all by eighteenth-century belief in the literal equivalence of life and sunlight.[32] Here "poesy" is inserted between life and light, strongly suggesting that creative energy is a power commutable with the other two.

This is stated more explicitly in stanza 163, which continues the description of the statue.

> And if it be Prometheus stole from Heaven
> The fire which we endure, it was repaid
> By him to whom the energy was given
> Which this poetic marble hath array'd
> With an eternal glory—which, if made
> By human hands, is not of human thought;
> And time himself hath hallowed it, nor laid
> One ringlet in the dust—nor hath it caught
> A tinge of years, but breathes the flame with which 'twas wrought. (*BW* 2:179)

In Aeschylus's account, Prometheus steals fire from heaven to provide man a tool and servant. But in another version of the myth, taken from Aesop and used by Goethe among others, Prometheus steals fire to animate the first men, whom he had molded out of clay. "The fire which we endure" suggests the second version—since fire is used here to figure life, understood as passion and as power. The salient point about this fire is its transferability; it circulates from hand to hand, changing form along the way.

Since Prometheus's theft is repaid by the creation of the Apollo Belvedere, solar fire seems to be interconvertible with life and with the "energy" of artistic creation. But these are not powers that an individual subject can possess: the statue's glory "if made / By human hands, is not of human thought." The impersonality of power, in effect, is its divinity.

In his reading of this stanza, Harold Bloom explains its attention to the pains of creation as a myth of hubris, exploring the individual mind's guilty usurpation of a divine prerogative.[33] This is in keeping with a reading that focuses on the individual creator's Romantic strength, and it accurately captures one emotion the stanza expresses. But Bloom's reading does not account for the stanza's emphasis on the impersonal circulation of energy, which permits Byron to include his audience as fellow-sufferers when he describes "the fire which we endure." Artistic aspiration has a fatal grandeur in *Childe Harold*, but it comes across as grandeur (rather than self-indulgence) only because the author successfully implicates his audience in the aspiration. In particular, he draws on the scientific and medical ideas of his age in order to imply that the self-consuming fire of creation is a condensed and intensified version of normal mortal physiology.

The history of science cannot, of course, explain why Byron links self-consuming ambition to art in particular. In these stanzas, the myth of Prometheus may partly determine that choice. Byron's uneasiness about his own fame also contributes something. But since the drama of tormented creation appears in other contexts and in other Romantic-era writers, one needs to think about this theme in broader terms. It may be useful to reflect on the hope (articulated by Friedrich Schiller, but shared by many other thinkers in the period) that it would be possible "to restore by means of a higher Art the totality of our nature which the arts themselves have destroyed."[34] Byron was not as preoccupied as Schiller with the division of labor or the specialization of knowledge, but in the third canto of *Childe Harold* he certainly does define art as a response to a social world that has become excessively servile and complex—a response that paradoxically restores autonomy through heightened self-alienation. Byron's representation of art as a "fire which we endure" can thus perhaps be understood—on the broadest level—as an instance of the recurring Romantic drama that pits art and consciousness against themselves.

But those threads of discussion lead to other books, which have been written ably by other writers.[35] Here I would only like to add that the science of the Romantic era gave poets an image that vividly embodied this paradox, and hinted at a possible resolution. Since life in general was a self-consuming fire, Romantic writers who burned themselves up could expect to be received as examples of a shared human condition. This irony rooted in natural law allowed Byron to generalize his gloom as more than a deforming accident of biography. The image of life as self-consuming flame symbolically resolves inward struggle by condensing that struggle into an impersonal force shared by all creatures. It also suggests that the speaker's renunciation of "fame, ambition, strife" simultaneously lays claim to a different sort of distinction, measured along a simpler and more universal axis of power (*BW* 2:78). In Byron's writing that form of distinction cannot always be called culture; the dramatic masks he tries on are too various to be summed up that neatly. But there are certainly passages that do reify culture as abstract power—for instance, the invocation of "life, and poesy, and light" I have been discussing at the end of *Childe Harold*.

The condensation of poetry as empirical force, and especially as sunlight, helped an audience of readers who were not all poets see their own mortal fate in the poet's baffled aspiration. It also allowed writers to adapt Byronic autobiography to lives that were awkwardly deficient in public scandal. Finally, it provided an ambiguous symbol that could be used to (partly) Christianize Byron. For all these reasons, Byron's English imitators reproduced especially his protagonists' personal relationship with the sun. "Aspley Wood" (1819), by Jeremiah Holmes Wiffen (1792–1836), announces indebtedness to *Childe Harold's Pilgrimage* quite frankly in its preface. Byron used Mediterranean topography to discuss the celebrated dead associated with each spot; Wiffen proposes to tour Aspley Wood, the scene of his boyhood in Bedfordshire, in order to reawaken the personages who, for him, had become imaginatively linked with its sights and sounds—the heroes of Ossian, of Shakespeare, and above all, Childe Harold himself.[36] One salient divergence from Byron is that Wiffen is a Christian optimist, albeit one for whom religion takes mainly the form of a walk in "some religious wood, or dim romantic glen." Wiffen spends several pages trying to undo the "deep misanthropy" he loves in Byron by urging the author to seek consolation in religion.[37] Since a majority of Byron's admirers were Christians, this was a common, if quixotic, response to his work. The struggle between religion and Byronism staged in "Aspley Wood" seems to have more of an effect on Wiffen's system of thought, finally, than on Byron's. In particular, the materialistic, mournfully defiant relation to the sun borrowed from Byron wreaks havoc on the consistency of Wiffen's Christianity.

The poem begins with the sun rising over Aspley Wood. A long address to the sun echoes *Manfred*, and broaches the inevitable theme of mortality by linking the antiquity of sun-worship to the antiquity of the sun itself. The sun-worshippers have disappeared, and yet the sun remains, the "Cradle of infant Time—his womb, birth, being, end!" Partly by analogy to the disappearance of antique sun-worship, Wiffen translates the loss of Eden into the loss of a primordial bond between the sun and the earth.

He laments that loss by addressing the sun in these terms:

> And we became all shadow—in the abyss,
> The spirit's desolation, here we stand,
> Wrestling in darkness for a heavenly bliss,
> And an immortal's essence: brightly grand,
> How climbest thou thy skies? nor lend'st a hand
> To help us to thy altitude! away
> Earthborn repinings—ye may not command
> A sparkle of that intellectual ray,
> Which yet from heaven descends, and communes with our clay.[38]

As a lament for the loss of (solar) Eden, this is not especially contrite. It is in fact defiant: "away / Earthborn repinings" comes across as a gesture of magniloquent despair rather than humility. When God was compared to the sun in seventeenth- and eighteenth-century Christian poems, the point was that he does "lend a hand," beaming beneficently on mortals below. Here Wiffen has subordinated that old Christian metaphor to a personal relationship with something very different: the sun's light itself, understood as an "intellectual ray" that men and women depend on but can never finally possess. This passage could be described as a secularization of Christian motifs only by a critic determined to view the matter upside-down. The pathos of the passage is akin, not to Genesis, but to Davy's "Spinosist," or Byron's equivocal praise for the sun as "the kindler of this dark lanthorn of our external existence."[39] The passage is best described as an incomplete Christianization of a theme borrowed through Byron from natural philosophy.

"The Girl of Provence" (1823) by Bryan Procter Waller ("Barry Cornwall") provides another example of the way late-Romantic writers use the sun to translate artistic dilemmas into physiological ones. Eva, a young French woman "fed . . . / With radiant fable," dreams that she becomes Apollo's bride. Abandoned by the god, she pines, worships a statue of Apollo that she finds in a museum, goes mad, and dies. The poem takes as its epigraph three lines from Byron's description of the Apollo Belvedere in *Childe Harold*, and its conception of divinity, like Byron's "Sun in human limbs arrayed," treats deity as a figuration of the power of nature: in Apollo's statue, the Girl of Provence sees "no stone or mockery shape / But the resistless *Sun*."[40]

The poem is also concerned to make its heroine's madness an emblem of the poet's fate. Like a score of other late-Romantic poems, "The Girl of Provence" uses a mortal's love-affair with a god in order to dramatize a gulf between reality and the imagination—and more particularly, in order to show how an unrealizable ideal can at once poison and enchant mortal existence.

But in Waller's poem Apollo doesn't give up his role as the physical sun in order to become a purely visionary light. His alterity is not bodiless or elusively transcendent; on the contrary, the solid marble statue that Eva encounters in the Louvre *is* Apollo as far as the poem is concerned, full of "Life and the flush of Heaven, and youth divine." Apollo is terrifyingly alien because of his immortality: "the terrors of his face . . . shine / Right through the marble, which will never pine / To paleness though a thousand years have fled, / But looks above all fate, and mocks the dead." The cultural immortality of the statue is conflated with the deathlessness of the god, and both are conflated with the life-sustaining power of light itself. Apollo becomes "insufferable day"—a dazzling, mocking excess of the power that sustains life, and of "the lightning of the passions,—in whose ray / Eva's bright spirit wasted, day by day." The narrative is not a quest-romance, but a story of unrequited love, and the heroine doesn't pine away for unattainable transcendence so much as burn up in the heat of her own imagination. She dies on a summer morning, awakened and brought briefly back to sanity by "the light she loved" coming through the window; she is also consumed by it, dying "when no more she could endure."[41]

One of the poem's most interesting passages ascribes this self-destructive fate not just to poets, but to poetry itself.

> And beautiful is great Apollo's page:
> But they who dare to read his burning lines
> Go mad,—and ever after with blind rage
> Rave of the skiey secrets and bright signs:
> But all they tell is vain; for death entwines
> The struggling utterance, and the words expire
> Dumb,—self-consum'd, like some too furious fire.[42]

Poetry is defined here as self-consuming language; the same influence that awakens poetry also burns it up from within, "and the words expire/ Dumb." In the context of Eva's death, the image of "some too furious fire" strongly evokes the Brunonian theory that illnesses are caused when life feeds too rapidly on its own fuel. The naturalistic myth that Davy and Byron borrowed to define human aspiration is being used here to express the germ of a poetic theory. That theory is more familiar, and is expressed more explicitly as a theory of culture, in Percy Bysshe Shelley's "Defence of Poetry."

The Sun's Transforming Power: Shelley's "Hymn of Apollo"

Shelley's poetic theory was organized and in part determined by the solar metaphors he used to express it. The best way to show this is to approach

the "Defence of Poetry" (1821) through the "Hymn of Apollo" (1820),
where Shelley worked out some of the central metaphors of the later text.
Shelley borrows contemporary photochemical theories in order to envision
a radiance that reveals the world by reproducing the world within itself. In
the "Hymn," that complex figure is linked to poetry through Apollo; in the
"Defence," it becomes a theory of culture in general.

The "Hymn of Apollo" is one of a pair of poems that Shelley
contributed to Mary Shelley's play *Midas*, which was based on a story from
Ovid's *Metamorphoses*. Apollo and Pan are engaged in a singing contest that
the hill god Tmolus will judge. After both deities sing, Tmolus gives the
prize to Apollo, but the mortal king Midas prefers Pan's singing. Apollo, dis-
pleased, gives Midas the ears of an ass. In Ovid, the king's preference for Pan
is a sign of uncouth, rustic judgment, but Mary Shelley balances the scales
by letting Pan plausibly argue that Tmolus's choice reflects a vested interest:

> Old Grey-beard, you say false! you think by this
> To win Apollo and his sultry beams
> To thaw your snowy head, and to renew
> The worn out soil of your bare, ugly hill!⁴³

In fact, the songs complement each other, so that a final choice between them
is neither possible nor necessary. Earl Wasserman aptly calls Pan's hymn a celebration of "lived experience." But in describing Apollo as "the mind's ideal
and abstract powers" or "absolute subjectivity," Wasserman too confidently
humanizes the sun god to make the singing contest dramatize opposed possibilities of human existence.⁴⁴ The opposition is on a larger scale. Apollo's
power, of which the mind is a portion, extends beyond human capacities; it
can, for instance, renew the "worn out soil" of a "bare, ugly hill." It brings the
world to consciousness, both as light and as the power of mind, without
becoming itself a point of view or a power that a single mind could possess.

Apollo's "Hymn" accordingly speaks not about his perception, but about
his actions and effects. The second stanza stresses the immediacy of his
presence in the world he illuminates:

> Then I arise, and climbing Heaven's blue dome,
>   I walk over the mountains and the waves,
> Leaving my robe upon the ocean foam;
>   My footsteps pave the clouds with fire; the caves
> Are filled with my bright presence, and the air
> Leaves the green Earth to my embraces bare. (*SP* 612)

Writing "bright presence" rather than "light" allows the poem to forget for
the moment that the sun is a distant body sending out rays, in order to

define sunlight instead as a power that fills and constitutes transparent space. The sun does not pass through the air; rather, the air's transparency is a nakedness that leaves the earth "bare" to the sun's "embraces."

But the solar omnipresence that the second stanza represents as nakedness recurs in the fourth stanza, paradoxically, as clothing. Transforming things into incarnations of himself by robing them in his light, the sun also reveals them in the only possible way. Mediation of the world through sunlight is the only visual immediacy, and the sun's clothing is the only possible nakedness.

> I feed the clouds, the rainbows, and the flowers
>   With their aethereal colours; the moon's globe
> And the pure stars in their eternal bowers
> Are cinctured with my power as with a robe;
>   Whatever lamps on Earth or Heaven may shine
> Are portions of one power, which is mine.   (*SP* 613)

Since the theory of poetry Shelley developed in the following year also depends on the figure of a power that "transmutes all it touches" by clothing things in its own light, it is worth looking at this passage in some detail.

In writing, "I feed the clouds, the rainbows, and the flowers / With their aethereal colors," Shelley chooses examples that show (as "feed" implies) that the sun's light not only reveals but also constitutes color. The first two examples draw on the Newtonian idea that all colors are latent in sunlight and become visible when separated. Separation through refraction creates the rainbow. Clouds acquire their colors, according to Newton, through selective reflection: sufficiently small particles (though in themselves transparent) reflect different colors of sunlight depending on their size.[45] In referring to "flowers," Shelley may be drawing on the same Newtonian idea and merely pointing out that the colors of flowers are themselves reflected sunlight. But the rest of the stanza suggests that he is alluding to a more recent hypothesis: that the sun's light endows all living things with color through its chemical action. After Jan Ingenhousz discovered that plants carried out chemical reactions when exposed to sunlight, a number of writers, including the self-taught scientific lecturer Adam Walker (1731–1821), concluded that the color, fragrance, and combustibility of plants were effects of the sunlight or "elemental fire" stored within them.[46] Shelley heard Walker's lectures at Syon House and again at Eton.[47] Davy, a more widely respected source, also attributed the colors of plants and animals to sunlight and stated specifically that "flowers owe the variety of their hues to the influence of the solar beams" in *Elements of Chemical Philosophy*, which Percy Bysshe owned and read with Mary late in 1816.[48]

Apollo's claim to "feed" the earth with color thus probably draws on theories about the chemistry of light. Chemical theories are certainly invoked by the stanza's last two lines: "Whatever lamps on Earth or Heaven may shine / Are portions of one power, which is mine." The lamps of Heaven are easily explained: Shelley would have known that the moon and planets are seen by reflected sunlight, and that the fixed stars are globes like the sun, producing light by the same (unknown, but probably chemical) processes.[49] "Lamps on Earth" are portions of the sun's power because the combustible principle in wood, oil, and tallow was "fixed light," absorbed by plants and stored by the animals that eat them.[50] So if the sun "cincture[s]" the world with its power "as with a robe," that robe is much more than a superficial covering. The things lit by the sun are also literally portions of its own power.

But "lamps on Earth" can also include more than physical light. Apollo claims responsibility for moral light in the third stanza, and the poem ends by defining a new kind of light one could call "Poetry"—although Apollo's power is already so expansive that Raymond Williams's word "culture" would convey the intended meaning more precisely.[51]

> I am the eye with which the Universe
>   Beholds itself and knows itself divine;
> All harmony of instrument or verse,
> All prophecy, all medicine is mine,
>   All light of art or nature;—to my song
> Victory and praise in its own right belong. (SP 613)

This stanza expresses the Romantic reification of poetry as power in its most complete and most articulate form. Here, as throughout the poem, Apollo is both a natural force and a power of the mind; he encompasses "all light of art or nature."[52] Apollo's power permits self-knowledge—but he is not himself a subjectivity so much as an "eye" through which the Universe comes to collective awareness. Apollo's eye is a peculiar lens, however: one that transfigures what it sees. The Universe doesn't just behold itself through Apollo; it "knows itself divine." It knows itself, that is to say, by recreating itself in a diviner form, just as Apollo's "bright presence" lays the world bare only by robing it with color. Finally, Shelley playfully stresses Apollo's mythic responsibility for prophecy and medicine, in order to suggest—as he will argue at more length in the "Defence"—that the power of poetry pervades all human culture.

Shelley himself may well have understood visible and poetic light to be unified on an ideal plane. In a world where "nothing exists but as it is perceived," light is no more material than poetry; both are aspects of a larger power that constitutes all perception (*SPr* 174). But Shelley's idealism changes nothing essential about the solar figure constructed in Apollo's "Hymn." It doesn't matter for the purposes of the poem whether sunlight and poetry are unified (as in Davy and Byron) because the mind's energy is understood to be physically derived from the sun, or whether the two are unified because sunlight is taken as a part of an ideal power also revealed in poetry. The poem is more interested in exploring the paradox of a power that reveals things by recreating them as part of itself. Whether understood in idealist or materialist terms, this model is indebted to eighteenth-century science: it is a version of Jean Senebier's fascination with light's double role as the spectacle of nature and the power behind the spectacle.[53]

Shelley's "Hymn of Apollo" importantly diverges from earlier treatments of the same theme not in its idealist ontology but in its radical skepticism. This skepticism surfaces in Apollo's remark that the moral effect of his light lasts only "until diminished by the reign of Night." The stress on transience suggests that the absolute illumination represented by Apollo is always qualified, always already fading, in the temporal realm of human experience. The "Hymn of Pan," exploring that realm, emphasizes Apollo's limitation by dwelling on twilight:

> Liquid Peneus was flowing
>   And all dark Tempe lay
> In Pelion's shadow, outgrowing
>   The light of the dying day,
>   Speeded by my sweet pipings. (*SP* 613)

Present participles—"flowing," "outgrowing," and "dying"—stress that twilight is not a moment but a process. The sun enters the mutable world only as a "shadow" or as something already in decline. Pan's skepticism about Apollo's light also takes the form of knowing bitterness about poetic power. The poetic "reed" breaks the moment Pan clasps it, and wounds his breast: "Gods and men, we are all deluded thus!" Together, the hymns of Pan and Apollo frame a complex elegiac statement of dynamic naturalism: all phenomena are "portions of one power," but that power can never be apprehended as a stable presence. Whether it is understood as sunlight or as poetry, it always contains within itself the seeds of its own destruction.

The Sun of Culture in the "Defence of Poetry"

The solar imagery worked out in Shelley's "Hymn of Apollo" reappeared a year later in the "Defence of Poetry." The "Defence" seeks, even more systematically than the "Hymn," to expand poetry into an abstract power that animates all human culture. Shelley achieves this expansion largely through metaphors borrowed from the science of light. Optical science gives him a way to explain how a mimetic power (a power that imitates and colors) can also be a power that permits true knowledge. The chemistry of sunlight allows him to explain how the same mimetic power can poignantly embody the self-consuming energies of human life.

Shelley first planned the "Defence" as a letter to the editor of *Ollier's Literary Miscellany*, in which Peacock's "Four Ages of Poetry" had appeared. He soon found that "the subject . . . requires more words than I expected" and composed a freestanding essay, but the first draft of the letter to Charles Ollier shows how the solar imagery he had elaborated in the "Hymn of Apollo" entered into his conception of the "Defence."[54] After a long, wry summary of Peacock's repudiation of poetry, Shelley writes, "these are indeed high objects [*& I pledge myself to worship Themis rather than Apollo if . . . if it could be found that. . . .*]" The draft then breaks off and begins again; the mention of Apollo has brought the sun to mind. "He would extinguish Imagination which is the Sun of life, & grope his way by the cold & uncertain & borrowed light of the Moon which he calls Reason,— stumbling over the interlunar chasm of time where she deserts us, and an owl, rather than an eagle, stare with dazzled eyes on the watery orb which is the Queen of his pale Heaven." The plan of Shelley's emerging argument, in short, is that Imagination is to life what the sun is to the visible world, and thus Peacock's idolatry of reason is merely a way of borrowing Imagination in a weak and reflected form. A few lines later, the strategy continues with the mock-outraged remark "[*I hope soon to see a treatise against the light of the Sun in . . . one of your columns*]."[55] But it must have been difficult to make these abstract metaphors as urbane and brief as a letter to the editor—and especially a polemical reply to a friend—demanded. The final version of the "Defence" becomes an essay rather than a letter; Shelley wisely drops all reference to Peacock; and the mock-outrage gives way to a measured analytical tone. But the underlying analogy—poetry is to life what the sun is to the visible world—remains the same.

The "Defence" repeatedly compares Poetry to light (*SPr* 276, 280, 282, 286, 294); more importantly, it borrows the conception of light developed in the "Hymn of Apollo"—light as a power that reveals the world by transforming the world into itself. "[P]oetry lifts the veil from the hidden beauty of the world and makes familiar objects be as if they were not familiar; it reproduces all that it represents, and the impersonations clothed in its Elysian light stand thenceforward in the minds of those who have once contemplated them as memorials of that gentle and exalted content which extends itself over all thoughts and actions with which it coexists" (*SPr* 282). This sentence's two halves, although connected by a semicolon, appear to make two contradictory claims. Poetry unveils the world, but it does so by impersonating the world and reproducing it in a newly clothed form. This business of taking clothes off by putting them on has puzzled commentators ever since M. H. Abrams rightly concluded that it exposes "a combination of Platonism and of psychological empiricism, and of the mimetic and expressive point of view."[56] The solution to the paradox lies not in Plato or Hume, however, but in ideas about the sun that Shelley drew from natural philosophy and had recently invoked in the "Hymn of Apollo." These claims for poetry are modeled on the Hymn's claim for the sun, whose light reveals things in their nakedness by clothing them "with my power as with a robe."

The "Defence" argues both that poetry creates new things never seen before, and that it makes perception and knowledge possible (*SPr* 293). It therefore needs both halves of this apparently paradoxical figure. If poetry were simply a way of clearing and aiding vision, it would be veridical but not creative. If, on the other hand, it were simply a transmuting power, it might seem to disguise reality rather than reveal it. What Shelley implies, therefore, is that poetry unveils things by transmuting them into itself. "Poetry turns all things to loveliness. . . . It transmutes all that it touches, and every form moving within the radiance of its presence is changed by wondrous sympathy to an incarnation of the spirit which it breathes; its secret alchemy turns to potable gold the poisonous waters that flow from death through life; it strips the veil of familiarity of the world and lays bare the naked and sleeping beauty, which is the spirit of its forms" (*SPr* 295).

The link between the two halves of the argument is sometimes hard to grasp in the "Defence," because the optical logic is not made as explicit as in the "Hymn of Apollo." But a solar figure is implicit in "the radiance of its presence," which encompasses the world as Apollo's "bright presence" did in the "Hymn." The rivers that run "potable gold" also echo the description of the sun in book three of *Paradise Lost*. There, the sun's "arch-chemic" power, compared to the philosopher's stone, makes rivers on the sun's surface run "potable gold."[57] Shelley's reference to a "secret alchemy" underlines the aptness of the allusion. Poetry for Shelley, like the sun for Milton, has a transmuting power. But the allusion serves ends that are determined by the poetic and photochemical discourse of Shelley's own time: nothing in Milton or in alchemical tradition suggested that the sun revealed the world *by* transmuting it. Shelley's insistence on the inseparability of the two processes develops out of his own poetic engagement with natural philosophy; it can be traced not to alchemy but to late-eighteenth-century chemists' fascination with the sun's ability to animate and enlighten simultaneously.

"Poetry" means for Shelley something very different from the archive of existing and potential poems. He defines "poetry in the universal sense" as a synthetic power latent in all perception that "perceives the before unapprehended relations of things" and organizes them into new forms. And he draws out the social consequences of this claim more boldly than most other Romantic theorists. If poetry is such a general power, the class of "poets" should include not only authors of verse but also "the authors of language and of music, of the dance, and architecture, and statuary and painting: they are the institutors of laws and the founders of civil society and the inventors of the arts of life," for "language, form, color and religious and civil habits of action . . . may be called poetry by that figure of speech which considers the effect as a synonym of the cause" (*SPr* 279). Poetry includes, in short, the whole realm that the nineteenth century would eventually define as "culture."

At the conclusion of the text, Shelley reminds readers that the "Defence" is intended as the first part of a two-part project. At the end of this first part Shelley takes himself to have established poetry's claim to universality by showing that "poetry in a restricted sense"—the activity of poets—is only an instance of the larger sun-like power he calls "poetry in a universal sense." The unwritten second part was to have been a more explicitly social manifesto: "a defence of the attempt to idealize the modern forms of manners and opinions and compel them into a subordination to the imaginative and creative faculty" (*SPr* 296). It would have argued, in other words, that the cloud of mind should discharge its accumulated lightning, and "the unacknowledged legislators of the world" should become acknowledged legislators of manners.

In fact, Shelley has already accomplished all this in the "Defence"; no sequel is needed. Instead of vindicating verse composition against Peacock's argument, he advances the claims of a larger power that shapes language, music, religion, law, and civil society. He calls that power poetry, but the more precise word would be "culture." Shelley asserts, in effect, that culture is a power deeply rooted in the mind; it has shaped all the forms of historical existence, and ought to be recognized as the true foundation of social distinction (compelling "the modern forms of manners . . . into a subordination"). This could easily sound like a plan for an artistic despotism; the phrase "legislators of the world" makes that aspect of the argument all too clear, and commonly alienates modern readers. Shelley's hope of evading that reaction rests on the universality of poetry (or culture). This is why it matters that poetry should be represented as a diffuse impersonal force rather than a personal acquirement. The identification of poetry with light becomes central to Shelley's argument because it helps him suggest that poetry is latent—not in poets, or cultural producers, or even in a "cultural class" of educated consumers—but in all perception.

It is also part of the strategy of the solar metaphor that poetry's presence should be fleeting and imperfect. The skepticism articulated in the "Hymn of Pan" reappears in the "Defence" as a tension between the claim that poetry is "the perfect and consummate surface and bloom of all things" and the claim that it is an "unapprehended inspiration" (*SPr* 297). Just as Apollo's light is always already "dying" by the time it enters the realm of Pan, "the mind in creation is as a fading coal," so that "the most glorious poetry that has been communicated to the world is probably a feeble shadow of the original conceptions of the poet" (*SPr* 294). This skepticism, aptly described by Angela Leighton as a refusal "to affirm Poetry as an abiding presence," is Shelley's way of creating the elegiac tension that gave Romantic solar myth its emotional power.[58] That myth had never celebrated stable presence; it represented identity as an ephemeral effect produced by a play of formless forces. In Humphry Davy's words, "No sameness and no deep identity, / Linked to the whole, the human mind displays." The mind, like any other flame, is only a "renovated form" of sunlight; its actions reflect a power located in the universality of nature, not a stable identity of its own.

The fading-coal metaphor in the "Defence" is thus more than a story about inspiration; it skillfully borrows pathos from medical and scientific thought, and uses that emotion to prove that culture runs as broad and deep as life itself. Brunonian physicians and philosophers had argued that life was a coal drawn from the sun, fanned by an inconstant wind, and consumed by its own brightness. Poets from Davy through Byron transformed that naturalistic argument into a myth of self-consuming genius. Shelley's fiery metaphor for the creative mind evokes the same story in order to remind readers that culture is an impersonal and universal force. Poets don't produce or authorize their own power, but rather participate in it, as all living creatures participate in vital power drawn from the sun. The effect of the analogy is both to underline culture's claim to universality, and to invite readers to see their own aspirations and mortal fate modeled in the activity of cultural production.

CHAPTER 7

ENERGY BECOMES LABOR: THE ROLE
OF ENGINEERING THEORY

In 1833 John Herschel wrote that "the sun's rays are the ultimate source of almost every motion which takes place on the earth," including the labor of human bodies, of falling water and of "those great deposits of dynamical efficiency which are laid up for human use in our coal strata."[1] In the same year Thomas Carlyle's professor Teufelsdröckh echoed the observation by remarking that the lowly "smithy-fire" glowing before him "was (primarily) kindled at the Sun"—a link displaying secret affinities between "Iron Force, and Coal Force, and the far stronger Force of Man."[2] As earlier chapters of this book have indicated, the connection between sunlight and human enterprise was not in itself new. Writers like Humphry Davy and Mary Robinson had foreshadowed Carlyle's fascination with a connection between sunlight and the "Force of Man" as early as the 1790s. But the tone of the fascination does change around 1830, and the change is at least as important as the continuity. When Mary Robinson in 1798 represents the sun as "the source refin'd / Of human energy," it is not quite the same thing as saying, with Herschel, that the sun is a source of "dynamical efficiency" (*RP* 3:1).

A received model of the Romantic era as a period of idealism makes it tempting to describe the difference by saying that Romantic connections between sunlight and energy are psychological, whereas Victorian writers interpret the connection physically. That would be a misleading analysis. Humphry Davy and Antoine Lavoisier had specific chemical reasons for believing that sunlight was responsible for human motion and sensation. Even in less scientific contexts, the word *energy* clearly included the activity of the body. William Blake remarked that "Energy is the only life and is from the Body"; William Godwin advised the man who "would have great energy" to "exercise and strengthen the muscles of every part of his frame" (*BP* 34, *GW* 5:185).

Yet it is correct to feel that Romantic exaltations of energy fail to encompass all forms of labor. The obstacle isn't a distinction between the body and the mind, but between categories that Hannah Arendt would have called "labor" and "work." According to Arendt, "the work of the hand" is anciently understood to be free and productive, whereas "the labor of the body" is identified with constraint and repetition.[3] Although Arendt approaches these concepts as primordial phenomenological categories, they manifest themselves in eighteenth- and nineteenth-century Britain above all as markers of social class. An artisan, a shopkeeper, or a teacher devotes free energies to work; a laborer, on the other hand, does "drudgery" or "mechanical and daily labor."[4] The underlying distinction is one of constraint: working-class labor is understood to lack the principle of autonomy that makes middle-class work energetic.[5]

In practice, the opposition between labor and work distinguished groups of people, rather than kinds of tasks. The proprietors of small farms performed many of the same daily tasks as agricultural laborers—but their work was not thought to constitute mechanic toil. What distinguished energetic work from toilsome labor, in the middle-class imagination, was "independence"—or, in other words, the fact that it was carried out by a person who controlled some portion of the means of production. According to William Godwin, "he who lives upon the kindness of another, must always have a greater or less portion of a servile spirit . . . True energy, the self-conscious dignity of the man, who thinks not of himself otherwise than he ought to think . . . are sentiments to which he is a stranger" (GW 5:183). So although a servant or a laborer might be diligent, they could not properly, in Godwin's view, be called energetic. The distinction is an ideological one: it interprets property as a boundary between two different kinds of human agency and thereby naturalizes social privilege. But the political effects of this distinction were not uniformly conservative; for Godwin, at least, the point was that society should be reorganized to reduce "mechanical and daily labour" and to guarantee the self-determination required for a full expression of human energy.[6]

To say that the sun provided human energy was thus not to say that it powered labor or toil. The categories were antithetical. Since late-eighteenth-century writers tended to assume that a state of nature was a state of liberty, it made sense to link the sun with free energy. Toil, as a form of constraint, was presumably unnatural, and Romantic writers did not equate it with natural force. In Thomas Love Peacock's fragmentary epic "Ahrimanes," for instance, the sun god Oromazes is identified with a prelapsarian state of natural production, whereas toil is assigned to Ahrimanes, god of night and evil artifice.[7] But by the end of the 1820s these distinctions between spontaneous work and toil, energy and mere diligence, began to collapse and give way to a generalized endorsement of productive work.

Many factors were involved in the shift. Dror Wahrman points out, for instance, that British writers became increasingly cynical in the 1820s about the ideal of economic independence. In the 1790s, writers like Godwin who celebrated independence thought it described a station of life with closely interrelated economic, political, and spiritual aspects. By the early 1820s, as the middle classes were enlisted as a bulwark against insurrection, popular satirists and Whig leaders alike were forced to recognize that middle-class property was not actually a guarantee of political independence.[8] These changes may have helped to erode the distinction between "energy" and "mechanic toil," which had hinged on confidence in the moral significance of economic autonomy.

But a more direct reason for the collapse of that distinction was the growing importance ascribed to flatly quantitative measurements of labor power. Between 1800 and 1817, as I mentioned in chapter 5, political economists sympathetic to manufacturing tried to show that nature contributed the same kind of surplus labor to industry that it was thought to contribute to agriculture. The elasticity of steam, and the strength of steel, they argued, were natural powers that contributed to production in the same way as the fertility of the soil. But as long as the claim that nature "works" in manufacturing was made in this abstract way, it remained remote from the distinction between different kinds of human work. The next step depended on quantification. Once engineers developed a general theory of machines as systems for accumulating and transforming motive power, it became possible to measure nature's contribution to production in the same units that were used to measure the contribution of human laborers. The new commensurability of productive agents was quickly drawn into the economic debate over the relative "naturalness" of agriculture and manufactures. Between 1829 and 1834, advocates for manufacturing interests seized on the correlation between human and machine labor and used it to speculate that human labor was not only commensurable with, but actually derived from, the motive power contained in sunlight. It thus became possible to represent the borrowing of natural power in water-powered and steam-powered engines as a paradigm for all economic production.

Once this happened, the old distinction between natural energy and mechanical toil became untenable. The characterization of working-class labor as mechanic (both manual and machine-like) no longer made sense once the overshot waterwheel and the steam engine were taken as models of natural production. By 1829, writers like Charles Babbage and G. G. Coriolis were arguing that all of nature's productive power could be measured in the same units that would be used to measure the output of a machine: weight lifted through a given height. As this realization spread in the 1830s, writers stopped making a systematic distinction between energy and mere toil. The old class distinction was replaced by a new distinction running between two categories of machines: those called "engines" or "motors," which draw their own motive power directly from nature, and other machines that merely apply and transform the power produced by a motor. A fantasy of autonomous power survived in this distinction, but it survived in terms that were socially much more ambiguous.

The dissolution of the boundary between "energy" and "toil" can be assessed from several different perspectives. I have stressed that the original boundary was middle-class ideology, in the classic sense that it embodied the middle classes' idealized view of their own position in the economy. The new quantitative description of labor did not express the self-conception of a particular class quite so immediately. But it was not devoid of political significance. In practice the quantification of labor power lent itself to an imperative that one can call "productivism," borrowing a word from the Belgian industrialist and philanthropist Ernest Solvay. The continuity between labor and natural force seemed to imply that the goal of all human endeavor was to divert more and more natural power to human ends, and therefore "that always and before everything else maximum production is the eminently moral supreme purpose to be pursued."[9] This normative claim, which Solvay named "productivism," idealizes not a particular class but industry itself, by making the diversion of power in water-powered or steam-powered engines a model for all human activity.

The social consequences of productivism emerge in the later nineteenth century. I do not have space here to offer a fresh interpretation of late-nineteenth-century social history, but I do want to briefly explain the relatively positive tone I take in surveying the early stages of this process, since readers in my own discipline of literary studies have traditionally taken a darker view. When literary scholars write about industrialization, they often adopt a perspective like the one Georg Lukács adopted in *History and Class Consciousness* (1922)—a perspective that traces the deformation of human life under capitalism back to the reification of labor power (or indeed, to reification as such).[10] I share the concern, but I do not trust the historical analysis. Quantification certainly objectifies labor-power and—in that very abstract sense—alienates it from the worker. But it might be a form of historical idealism to conclude, reasoning backward, that the actual economic exploitation of workers was fostered by this act of abstraction. Anson Rabinbach's study of late-nineteenth-century sciences of work, *The Human Motor*, argues that the quantification of labor power in fact provided a justification for important progressive reforms, such as the institution of a ten-hour day.[11] The quantification of labor power figures centrally, for that matter, in the later writings of Marx and Engels.[12] A generalized critique of "reification" or "commodification" can too hastily equate intellectual and political history. To isolate a dimension of human life for analysis is not necessarily to subject that dimension of life to unregulated market forces.

My own belief—although I do not pretend to have proved it decisively in this chapter—is that the quantification of labor power marks a small intellectual and moral advance over the older, crudely ideological distinction between energy and mechanic toil. Quantification is no guarantee of objectivity, but quantification is one way to test the unexamined, self-interested, and wishful assumptions that pervade human reasoning. That testing process does not deliver a value-free world—nor would one want it to.[13] But it may nevertheless promote clearer thinking about the—admittedly blurry—boundary between facts and values. Early-nineteenth-century engineers were not neutral observers; the quantitative description of work they developed certainly did reflect their own investment in industrial capitalism. Quantification could be applied heartlessly. But it provided a clearer and more candid description of work than had yet been articulated—and over the long run, if Rabinbach is correct, that clarity also advanced the interests of people who were not engineers.

The Scarcity of Labor

In chapter 5, I discussed Jean-Baptiste Say and David Ricardo together as political economists who stressed that manufacturing depends on nature's assistance. But the theories of production they advanced differ significantly. Say represents "the work of nature" as a real contribution to the value of products. Ricardo agrees with Say that "we are almost always working in concurrence with nature," but also recognizes that "nature" defined this broadly becomes an empty category.[14] His recognition takes the form of a distinction between use value and exchange value. Sunlight and air are useful, but because they are available gratis, no one will give anything for them. They therefore add nothing to the exchange value of the commodities they help to produce. According to Ricardo, only one agency actually augments the exchange value of commodities—human labor. Adam Smith had thought that the cost of rent and capital formed part of the price of commodities as well, but Ricardo analyzes those costs so as to resolve them into labor. Farmers pay rent for the use of productive and conveniently situated land. But what this really means, for Ricardo, is that it requires more labor and capital to produce a given amount of wheat on one parcel of land than it would require to produce the same amount on another. Rent thus resolves into the cost of labor and capital. And everything that capital adds to exchange value can be resolved into the cost of the labor that initially produced that capital, spread out over time. The rent paid to a landlord can in the end be explained as the difference between the labor cost of production on his land, and the higher labor cost of production on the poorest land currently under cultivation. Say's triad of productive factors—natural agents, labor, and capital—all reduce for Ricardo to human labor.

Ricardo, then, is not a productivist. He argues that "natural agents" play an essential role in all production, but add nothing to exchange value because the labor of natural agents is not scarce. This conclusion was inevitable so long as nature was conceived as a collection of discrete "agents": specific laws (gravity), or objects (steel). An invariable property—gravity, or the strength of steel—is not scarce, and has no exchange value. But other disciplines were defining ways of thinking about this question that did not presuppose the indivisibility of agency. French engineers, in particular, had already developed a general theory of motors that represented machines and human bodies as systems for accumulating and transforming motive power. The theory emerged in the late eighteenth century, but its consequences for political economy did not become clear until 1829, when Charles Babbage in England, and G. G. Coriolis in France, both published texts arguing that nature's motive power is a scarce commodity with economic value. With the addition of this premise, Ricardo's labor theory of value became productivism: labor remained the source of all value, but nature became the source of all labor.

In the simple form of a product of weight and height, the measurement that Coriolis would eventually define as "work" had been discussed under various names—"effect," "duty," "mechanic power"—since the middle of the eighteenth century. It had acquired considerable importance for practical mechanics, first in connection with water power, and later in connection with steam. The mathematics of its relation to other dimensions of movement (in particular, to the quantity Leibniz had called *vis viva*) were well understood by the 1770s. What was not widely recognized—not until the late 1820s in Britain, and only slightly earlier in France—was its economic significance as a measurement of the labor of machines, animals, and human workers.

The reasons for this lag between practice and theory were social rather than technical, and have to do mainly with a paucity of interest in the mathematical description of machines as productive moving systems.[15] Working mechanics were interested in improving the efficiency of machines, but not (by and large) in defining what machines essentially do in general mathematical terms. The mathematicians and natural philosophers who wrote treatises of rational mechanics, on the other hand, were interested in machines mainly as models for the equilibrium existing between forces external to the machine—not as moving systems that overcame resistance. The discourses of practical and rational mechanics were not in principle incompatible, and exchanges certainly did take place between them. John Smeaton, for instance, contributed articles to *Philosophical Transactions* on a mathematical theory of "mechanic power," and his observations had considerable practical influence. But these exchanges lacked a lasting social basis, because there were no institutions in late-eighteenth-century Britain to provide educational support, or professional rewards, for the systematic mathematical investigation of engineering problems. It was therefore in France, where the first institutions for the mathematical training of civil and military engineers took shape, that machines were first described generally as systems for the conservation and transformation of motive power.

There were also general philosophical reasons why writers were not inclined to see the product of weight and height, or force and distance, as a particularly significant quantity. I have argued that late-eighteenth-century culture was characterized by a heuristic that strove to abstract agency as such from the objects and events that instantiated it. But writers made this effort of abstraction without giving up the notion that agency was something possessed (in principle) by an agent. Even while attempting to trace the circulation of agency, in other words, Romantic-era writers still understood agency in personal terms—as a kind of essence of agent*hood*. This assumption stands out especially clearly in mathematical discussions of motive power. In the first decade of the nineteenth century, British engineers still represented motive power as a property belonging to and uniquely characterizing the agent from which it springs. Given an agent, in other words, one ought to be able to specify the amount of "moving power" said agent can provide. In order to measure agents in this way, time must be factored out of the equation. Otherwise one would be in the odd situation, as one early-nineteenth-century writer put it, of claiming "that the force of a child is equal to that of a man carrying a load, because the child is also capable of carrying the same load, though in small parts and in a greater length of time."[16] In order to avoid this odd result, motive power was usually defined mathematically as the change (whether of position or of velocity) that an agent could produce in a fixed and arbitrarily small unit of time.[17] Moving time into the denominator made it possible to use these measurements to characterize agents—and gave motive power the appearance of an effort made "at will" rather than a scarce and exhaustible commodity.

Late-eighteenth-century engineers did realize that there were ways of measuring motive power that ignored time. It is instructive to compare John Smeaton's *Experimental Enquiry Concerning the Natural Powers of Water and Wind* (1759) to a paper he published revisiting the same topic in 1776. Smeaton was trying to predict the amount of work a given fall (or "head") of water could perform. The 1759 *Enquiry* measures this by measuring "the raising of a weight, relative to the height to which it can be raised in a given time."[18] The power of a given head of water could thus be represented as the product of the weight of water flowing through the sluice in a minute, and the total height of its fall. Because the distance a body falls under constant acceleration is proportional to the square of its final velocity, the product of weight and height is proportional to the *vis viva* of natural philosophy (mv^2), but Smeaton chose not to emphasize this consequence in 1759. In a paper of 1776, however, Smeaton revisited his technique for measuring power, and pointed out its connection to *vis viva*. Through characteristically painstaking experiments, he showed that a weight lifted to a given height would always produce the same *vis viva*, whether it fell freely, or descended slowly attached to a cord that spun a set of weighted arms. Time, therefore, could be bracketed. "A mechanical power . . . properly speaking, is measured by the whole of its mechanical effect produced, whether that effect is produced in a greater or a lesser time."[19] Smeaton's observations on waterwheels contributed to the widespread replacement of undershot wheels by overshot or breast wheels.[20] The dimension he measured in 1759—the height to which a given weight could be raised in a given time—was widely adopted as a way to compare different sources of mechanical power against each other. But the remark Smeaton made in 1776—that time may as well be left out of the reckoning, in which case this measurement is proportional to *vis viva*—was ignored, in British practice, well into the nineteenth century.

A main source of resistance to Smeaton's suggestion was the widespread assumption that mechanical power should be understood as a characteristic attribute of an agent. The article on "Mechanics" in the 1810 edition of the *Encyclopædia Britannica* provides a good illustration of the way this assumption continued to deform mathematical reasoning on the topic of motive power. The article includes, under the heading of "Practical Mechanics," a table comparing various "first movers" in terms of the height to which they can lift a weight in a given time. Each row of the table correlates a height with the size of various "first movers" that could lift a thousand pounds avoirdupois that high in a minute. The power provided by horses and men is measured in numbers of laborers; the power provided by water, wind, and steam is measured by the diameter of the wheel, sails, or engine cylinder. The qualification that power should be measured "in a minute" may seem like a relatively minor consideration. But in fact the insistence on specifying time is essential to the organization of the table; without it, there would be no way of establishing a unique measurement of power for each agent. One would have to compare man-hours, horse-hours, and bushels of coal, rather than men, horses, and engines. In reality, that would be a better measurement, since the diameter of a wheel or engine cylinder does not reliably correlate with power produced. What matters is the amount of heat, or water, made to flow through the engine. Smeaton's work had already made this

TABLE shewing the relative strength of Overshot Wheels, Steam Engines, Horses, Men, and Wind-mills of different kinds.

Number of ale gallons delivered on an overshot wheel, 10 feet in diameter, every minute.	Diameter of the cylinder in the common steam-engine, in inches.	Diameter of the cylinder of the improved steam-engine, in inches.	Number of horses working 12 hours per day, and moving at the rate of two miles per hour.	Number of men working 12 hours a-day.	Radius of Dutch sails in their common position, in feet.	Radius of Dutch sails in their best position, in feet.	Radius of Mr Smeaton's enlarged sails, in feet.	Height to which these different powers will raise 1000 pounds avoirdupois in a minute.
230	8.	6.12	1	5	21.24	17.89	15.65	13
390	9.5	7.8	2	10	30.04	25.30	22.13	26
528	10.5	8.2	3	15	36.80	30.98	27.11	39
660	11.5	8.8	4	20	42.48	35.78	31.30	52
790	12.5	9.35	5	25	47.50	40.00	35.00	65
970	.	10.55	6	30	52.03	43.82	38.34	78
1170	15.4	11.75	7	35	56.90	47.33	41.41	90
1350	16.8	12.8	8	40	60.09	50.60	44.27	104
1435	17.3	13.6	9	45	63.73	53.66	46.96	117
1544	18.5	14.2	10	50	67.17	56.57	49.50	130
1740	19.4	14.8	11	55	70.46	59.33	51.91	143
1900	20.2	15.2	12	60	73.59	61.97	54.22	156
2100	21.	16.2	13	65	76.59	64.5	56.43	169
2300	22.	17.	14	70	79.49	66.94	58.57	182
2500	23.1	17.8	15	75	82.27	69.28	60.62	195
2686	23.9	18.3	16	80	84.97	71.55	62.61	208
2870	24.7	19.	17	85	87.07	73.32	64.16	221
3055	25.5	19.6	18	90	90.13	75.90	67.41	234
3240	26.25	20.1	19	95	92.60	77.98	68.23	247
3420	27.	20.7	20	100	95.00	80.00	70.00	260
3750	28.5	22.2	22	110	99.64	83.90	73.42	286
4000	29.8	23.	24	120	104.06	87.63	76.68	312
4460	31.1	23.9	26	130	108.32	91.22	79.81	338
4850	32.4	24.7	28	1.	112.20	94.66	82.82	304
5250	33.6	25.5	30	150	116.35	97.98	85.73	390

Figure 7.1 "Mechanics," *Encyclopaedia Britannica*, vol. 13 (Edinburgh: 1810), 121. Division of Rare and Manuscript Collections, Cornell University Library. Page 255.

clear, but the author of the table continues to rely, for rhetorical convenience, on the assumption that agency comes embodied in discrete agents.

There was less of this muddled thinking in France. The main reason for the national difference lies in the superiority of French institutions for engineering education—which is also to say that it lies in the longer persistence, in Britain, of a class-based segregation of intellectual disciplines. British natural philosophers were not primarily interested in the problem of measuring work. British engineers were not being educated in the calculus. There were no institutions in Britain comparable to the *École de génie* (founded 1748) or the *École des ponts et chausées* (founded 1747–1775), which provided rigorous academic training in mathematics for civil and military engineers in France.[21] There was a Corps of Royal Engineers, to be sure, but "their training was wholly practical, i.e. working in the royal dockyards of the Navy or at the ordnance department of the Army." The British School of Military Engineering was not established until 1812. Smeaton's Society of Civil Engineers, founded in 1771, remained "essentially a dining club."[22]

Even in France, the rigorous mathematical training of engineers was a recent development. It is probably not an accident that Lazare Carnot (1753–1823), who belonged to the first generation of engineers to receive such training, was also the first writer to propose a general theory of machines in terms of the product of force and distance. Carnot's *Essai sur les machines en général* (1783) can be seen in retrospect as inaugurating a genre of texts on "the general theory of machines" that would lead eventually to the work of Babbage and Coriolis. But Carnot's *Essai* did not win the prize for which it was twice submitted, and it seems to have received little attention until a revised edition was published under a different title in 1803.[23] By that time, Carnot had become well known for reasons that had little to do with his scientific work. From August 1793 to May 1795, Carnot was the supreme director of French military strategy. The only close colleague of Robespierre to continue in office after Robespierre's fall, he was a leader of the Directorate between November 1795 and September 1797 and served again (briefly) as Napoleon's minister of war. Carnot was thus somewhat better positioned in 1803 to find readers than he had been as a young engineering officer. But the potential audience for the study had also expanded. As Charles Gillispie points out, the audience for this sort of study would necessarily be an audience of professional engineers—an audience that had only recently been invented.[24]

Carnot argued that previous writers on mechanics had treated machines as if they were massless systems for transferring force between bodies outside the machine itself. While this approach works well for conditions of equilibrium, it leads to a misleading treatment of dynamic problems.

The concepts needed to explain the behavior of actual machines in motion are, on the one hand, the product of force and distance (which Carnot calls "moment d'activité," or sometimes "force vive latente") and on the other hand, "force vive réelle," "live force," which is simply Carnot's translation of *vis viva*. These terms are equivalent to the present-day terms "work" and "kinetic energy,"—with the proviso that Carnot is still using Leibniz's measure of mv^2 for *vis viva*, rather than $\frac{1}{2}mv^2$. Carnot's "moment d'activité" can be expressed as the product of a weight p and a height h. And since, according to Galileo's law of falling bodies, $ph = \frac{1}{2}mv^2$ this quantity is also proportional to the "force vive" a body of weight p would acquire in falling from that height. By describing an abstract ideal machine considered as a set of centers-of-gravity, Carnot is able to show that it is possible to measure the effect of all machines, and the power of their first movers, in terms of "force vive."

> [O]ne can always compare this effect to a weight to be lifted to a certain height, and by consequence to a *force vive*, either real or latent. Thus, for instance, to compress a spring a given amount . . . to reduce a given quantity of wheat to flour, to drive a carriage of a given load from a given place to another on a road whose roughness is given . . . are so many effects to be evaluated in terms of *force vive*.[25]

In this passage Carnot begins to draw attention to the economic significance of *vis viva*—an insight that subsequent writers would develop more and more explicitly.

Carnot already hypothesized, in 1803, an extension of *vis viva* to encompass the forces within animal bodies. To do so it was not necessary to understand the chemistry or thermodynamics of respiration; it was only necessary to consider the animal body from the point of view of its capacity to do work.

> It would thus seem that one could consider an animal, for physical purposes, as an assembly of particles separated by springs that are more or less compressed, which thereby conceals a certain quantity of live force. These springs in expanding convert latent live force into real live force. The greater or lesser quantity of latent live force in an animal is, strictly speaking, what one means by its degree of strength, when one says that this or that animal is stronger than this or that other one, and can produce a double, triple, quadruple, &c. effect.[26]

The hypothesis of springs is not to be taken literally; Carnot's point is merely that the animal body can be treated, for functional purposes, as an entity that stores live force in some latent form, and is capable of releasing it by converting it into "force vive réelle."

One of the first texts to follow in the genre Carnot had initiated was J. N. P. Hachette's *Traité Élémentaire des Machines* (1811). The first edition of this book in fact includes an approving "Rapport fait par M. Carnot" that underlines the filial relationship by pointing out that Hachette's phrase "effet dynamique" (measured as the product of weight and height) is equivalent to Carnot's own "moment d'activité." Hachette's book adds little of a mathematical nature to Carnot's theory, but it goes somewhat further than Carnot had gone in defining an abstract machine as something that accumulates and converts the "dynamic effects" produced by sources of motive power. Hachette, following Carnot, groups these sources together under the general term *moteurs*.

> Machines are moved either by animals, or by water, or by air, or finally by the action of caloric; each of these bodies is capable of producing movement, and for that reason, we call them *moteurs*. To compare different *moteurs* against each other, we measure the dynamic effect they produce in a given time. Of all dynamic effects, the simplest is the elevation of a weight to a certain height taken as unity—for instance, of a kilogram to a meter in height.

Hachette goes on to explain that the purpose of machines is not to generate power, but merely "to render a given *moteur* capable of a given dynamic effect," by converting weight into height, height into time, or time into weight. "Machines considered under this point of view are means of accumulating and conserving the forces that one or many *moteurs* produce in a given time, and employing them in another (greater or lesser) time."[27] Having defined "machine" so generally, Hachette is free to observe that hand tools are also machines, and machines are merely complex tools. In his introductory "Rapport," Lazare Carnot notes that a horse, too, is a "living machine."

I have left the word *moteur* untranslated for a reason. In the early nineteenth century neither French nor English had a word that meant precisely what "motor" means now—that is, a self-moving machine whose purpose is to convert energy from some latent form into motion. Lazare Carnot began to develop that concept when he modeled the animal body as a system of springs that "conceals" latent live force and can "convert" it into real live force. The distinction between motors and other machines, which would become important to the late-nineteenth-century economic imagination, appears first in engineering theory around 1810.

The Scarcity of Motive Power: Hachette, Coriolis, Babbage

From the perspective of political economy, the most important contribution Hachette made was to begin to calculate the economic value of "effet dynamique." For instance, "it is easy to calculate that the Chaillot machine does, in twenty-four hours, the work of 2058 water-carriers for 849 francs, which puts a man's day at the moderate price of 0,49 francs, and that of a horse seven times stronger than a man at 2,87 francs."[28] This less than compassionate reflection on the economic fate of manual labor is the sort of gesture that would soon call forth satirical attacks on calculation itself. The significance of these particular calculations, however, lay in Hachette's recognition that motive force was a commodity with measurable exchange value. The rapprochement between mechanics and political economy that Hachette began to suggest in this ad hoc fashion was given lucid theoretical expression by G. G. Coriolis (1792–1843), in *Du calcul de l'effet des machines* (1829). In the course of the second and third decades of the nineteenth century, the practice of treating machines as systems for the accumulation and conversion of "effet dynamique" or *vis viva* had become firmly institutionalized in French engineering education. Coriolis claimed to do little more than systematize and simplify this existing practice, but some of the changes he made in the name of clarity actually have far-reaching consequences. Foremost among these changes is a matter of terminology. In place of the variety of terms then in existence for the product of force and distance—"moment d'activité," "puissance mécanique," and "effet dynamique"—Coriolis suggests simply *travail*, or "work," a word that "seems to give a clear idea of the thing, entirely in keeping with its ordinary use in the sense of physical labor."[29]

Equally important for Coriolis, the word "work" underlines the economic significance of the measurement. In an important passage, Coriolis argues that work, the product of force and distance, is not only scarce, but *the* scarce commodity by which all production should be measured.

> We shall now show that . . . this quantity serves as the basis for evaluating *moteurs* in commerce; that it is work that one must seek to economize, and that all questions of efficiency in the employment of *moteurs* are related to this one quantity. We produce the goods that supply our needs in no other way but by displacing bodies or by changing their forms; this cannot be accomplished on the earth's surface except in overcoming resistance, and exerting certain efforts in the direction of movement. The ability to produce such displacement accompanied by force in the direction of displacement, that is to say the faculty we have called *work*, is therefore quite a useful thing. Whether

one draws it from animals, from water, or from air in motion, from the combustion of coal, or from the descent of heavy bodies, it is limited in every time and place; it cannot be created at will. Machines do nothing but employ and economize work, without being able to augment it; thus the faculty of producing work is sold, bought, and saved up like any other useful thing that is not found in extreme abundance.[30]

Coriolis was broadly educated, and this passage draws as heavily on political economy as it does on the tradition of French engineering mechanics. An important premise of its argument is borrowed directly from the work of J.-B. Say—that "production" is really a matter of moving objects into a useful position or changing their forms. To this Coriolis adds his own insight, that the mechanical capacity for producing these displacements and changes can be formalized mathematically as "work."

The economic consequences of this measurement, in turn, helped Coriolis establish its mathematical aptness. Coriolis uses the correlation between work and economic value to reinforce his decision to exclude time from the dimension of measurement.

Let us note also, that when it is necessary to produce a certain number of equal displacements with a machine, since in many cases it costs no more to bring them about simultaneously than it would successively, time cannot properly be brought in as an element of the value of this quantity of produced displacements. Suppose, for instance, that one proposes to employ ten men to lift certain weights: if one should for some reason desire to execute this task more promptly, one could always employ twenty men simultaneously; and though it would involve no greater cost in days, the same effect would be brought about in half the time.[31]

The economic assumptions here are not faultless: in assuming that a given number of man-days have the same cost whether spread out or bunched together on the calendar, Coriolis tacitly assumes that the supply of labor is unlimited, bypassing the very point he introduced earlier, that the supply of work "is limited in every time and place." But what Coriolis needs, and finds in the discourse of political economy, is a way of weaning the imagination from its accustomed focus on the capacities of discrete agents. By directing the reader's attention to measurements of economic value, Coriolis is able to show that his measure of work does have practical significance in the world, in spite of the unsettling fact that it cannot be used to uniquely characterize an agent. Coriolis's concept of work is thus both a contribution to political economy by the discourse of engineering mechanics, and an instance where economic considerations can be shown to have legitimized and partly determined engineers' choice of measurements and terminology.

British writers only gradually caught up with the French understanding of these matters. As late as 1813, for instance, Peter Ewart still found it necessary to suggest cautiously that *vis viva* might be an appropriate measure of moving force.[32] But it would be superfluous to detail the hesitant and piecemeal British acceptance of this principle. What matters here is less the mathematical point than the economic consequence of this line of reasoning—the realization that motive power is a scarce commodity that cannot be produced at will or drawn in unlimited quantities from nature. The British text that made that point most influentially was Charles Babbage's long article "On the General Principles which Regulate the Application of Machinery to Manufacture and the Mechanical Arts," which appeared in the *Encyclopedia Metropolitana* in 1829 and was reprinted separately as *The Economy of Machinery and Manufactures* in 1832. Babbage shows no sign of familiarity with Coriolis's book, published in the same year, but he duplicates some of Coriolis's economic conclusions. He reaches them, however, through a speculative and teleological argument rather than an explicitly mathematical one.

The organizing plan of Babbage's text is to sort machines into groups according to the fundamental "advantage" they provide in production: "Accumulating Power," "Regulating Power," or "Extending the Time of Action of Forces," for instance. The first and broadest of these classificatory divisions is the distinction between "*those* [machines] *which are employed to produce power, and . . . those which are intended merely to transmit power and execute work.*" This distinction (analogous to the French distinction between *moteurs* and other machines) brings up the question of what it really means for a machine to "produce power," and Babbage is quick to observe that in fact machines "neither add to nor diminish the quantity of motion in existence."[33] Babbage defends this claim by arguing that machines that apparently "produce" power form part of larger cycles in nature; since those cycles must (he assumes) be stable and constant, all the power that machines apparently produce must have been borrowed from a natural source.

Babbage discusses water power first, but for the sake of brevity it will be best to focus on his discussion of steam, where the logic is slightly clearer.

> The force of vapour is another fertile source of moving power, but even in this [case] it cannot be maintained that power is created. Water is converted into elastic vapour by the combustion of fuel. The chemical changes which take place are constantly increasing the atmosphere by large quantities of carbonic acid and other gases noxious to animal life. By what process Nature decomposes or reconverts these elements to a solid form, is not sufficiently known. The absorption in large quantities of one portion of them by vegetation is stated to take place; but if the end could be accomplished by

Mechanical force, it is probable the power necessary to produce it would at least equal that which was generated by the original combustion. Man, therefore, cannot create power; but, availing himself of his knowledge of Nature's mysteries, he applies his talents to diverting a small and limited portion of her energies to his own wants; and, whether he employs the regulated action of steam, or the more rapid and tremendous effects of gunpowder, he is only producing, in small quantity, compositions and decompositions which Nature is incessantly at work in reversing, for the restoration of that equilibrium, which we cannot doubt is constantly maintained throughout even the remotest limits of our system.[34]

This passage might be understood to foreshadow the doctrine of the conservation of energy, except that it has neither a clear definition of a conserved quantity, nor any evidence to defend its claims. Babbage takes a speculative approach that depends in large part on teleological assumptions about nature. Without stopping to define "moving power" or "mechanical force," Babbage blithely asserts that the mechanical force necessary to reverse combustion, *if* mechanical force could do so, "would at least equal that which was generated by the original combustion." Babbage's only basis for this claim is his model of Nature as a constant dynamic equilibrium, which rests (as the mention of "gases noxious to animal life" hints) largely on the assumption that the earth is designed as a stable home for human beings.[35] This assumption, speculative as it was, sufficed to convince Babbage that the power extracted from nature was a scarce and valuable commodity. He remarks, for instance, that the proximity of volcanic heat to glaciers in Iceland suggests the possibility of using the heat to compress, and the cold to liquefy, air—which could then be sold as a convenient source of motive power, so that "in a future age, *power* may become the staple commodity of the Icelanders. . . ."[36]

Political economists were quick to grasp the implications of this way of thinking. Nassau Senior's *Outline of the Science of Political Economy* (1836), originally written as the political economy article for the same *Encyclopedia Metropolitana* in which Babbage's article had appeared, refers readers to Babbage's article for an account of the economic function of machines, and supplements this reference with a long description of a particular case where power became a commodity: a mill on the Mersey, drawing power from the river, that rented rooms out to tenants who owned their own machinery and needed only a source of power. The point of the anecdote is explicitly to illustrate Babbage's systematic distinction between the machines that produce power and the machines that only transmit it to execute work.[37]

Senior proved to be a standard-bearer for a post-Ricardian school in British political economy. Along with Samuel Read, John Stuart Mill, and others, he moved away from Ricardo's insistence on labor as the sole source of value, back toward an updated version of J.-B. Say's triad of productive factors.[38] Senior recognizes, as Say had not, that things unlimited in supply cannot contribute to the exchange value of the commodities they help to produce. But he also recognizes, as Ricardo had not, that motive power is not unlimited, but a scarce commodity analogous to and commensurable with human labor.[39] Through the intervention of engineering theory, Ricardo's labor theory of value was thus transposed from the human realm to the realm of natural force, and became productivism.

The Contradictions of Productivism in Carlyle

Productivism is not merely the abstract belief that all labor comes from a common reservoir of motive power in nature; it is also an imperative to maximize production by consuming ever more power and deploying it ever more efficiently. The early writings of Thomas Carlyle make a strong case for the centrality of this imperative in early Victorian culture. "I too could now say to myself: Be no longer a Chaos, but a World, or even Worldkin. Produce! Produce! Were it but the pitifullest infinitesimal fraction of a Product, produce it in God's name!"[40] What is new about this injunction is not, of course, the work ethic itself, but rather the basis for Carlyle's endorsement, which is less Protestant than scientific: work restores the health of the soul (turns it from a "Chaos" into a "World") by putting it in contact with a great metabolism of productive force in nature. Work is sacred because it is the vital nexus through which human beings participate in a world that is itself productive. "Work, never so Mammonish, mean, *is* in communication with Nature."[41]

But the naturalistic exaltation of work, if it doesn't run back quite to John Calvin, was at least forty years old. Throughout the Romantic era, I have argued, the language of energy linked human production to the productive power of nature. What makes Carlyle's naturalistic argument productivist, and thereby distinguishes it from Romantic precursors, is the extension of Carlyle's naturalism to encompass working-class as well as middle-class production. Carlyle's mature writings do not distinguish autonomous, natural energy from mechanic toil. All work, be it "never so Mammonish, mean," is natural.

Yet this expanded definition of work wasn't felt by Carlyle as a liberation. His early writing began by attempting to set up a standard Romantic dichotomy between autonomous and constrained production, energy and toil; Carlyle's definition of work broadened only because this old distinction became incoherent in the context of an industrial economy. And even after Carlyle had abandoned the attempt to distinguish energy from toil, he remained troubled by the uncertain location of autonomy in the new productivist scheme. His writing on labor therefore complements its assertion of the unity of all "Force" with the countervailing claim that "Man's Force," in particular, should eventually encompass and command all others.

The best place to observe Carlyle's struggle to maintain Romantic dichotomies is the early essay "Signs of the Times," published in the *Edinburgh Review* in 1829. "It is," Carlyle writes, "the Age of Machinery, in every outward and inward sense of that word. . . . Nothing is now done directly, or by hand; all is by rule and calculated contrivance."[42] Under this general category of "calculated contrivance," Carlyle includes not only literal machines—power-looms and steam-engines—but Bible-societies, educational schemes, debates over parliamentary reform, and other sorts of "external combinations and arrangements" in society.[43] They are all signs of a focus on the external and mechanical rather than the primary and internal. Against these contrivances, Carlyle counterpoises what William Godwin and Humphry Davy would have called "the energies of my mind." Carlyle calls it "individual endeavour, and . . . natural force," but the adjectives in that phrase stress the same point that the word "energies" itself had concisely implied in the late eighteenth century—the spontaneous and autonomous nature of the power being exerted.

> To speak a little pedantically, there is a science of *Dynamics* in man's fortunes and nature, as well as of *Mechanics*. There is a science which treats of, and practically addresses, the primary, unmodified forces and energies of man, the mysterious springs of Love, and Fear, and Wonder, of Enthusiasm, Poetry, Religion, all of which have a truly vital and *infinite* character; as well as a science which practically addresses the finite, modified developments of these, when they take the shape of immediate "motives," as hope of reward, or as fear of punishment.[44]

The opposition between energy and artifice had been relatively simple for Romantic writers. On the one hand, natural force and spontaneous power; on the other, artifice and mechanic toil. In "Signs of the Times" Carlyle tries to set up the same opposition, but runs into a problem. The relation between the two poles, now defined symbolically as "Dynamics" and "Mechanics," has become too close. The distinction between "primary, unmodified forces" and their "finite, modified developments" begins to look like hair-splitting, if "Mechanics" is really only the form taken by "Dynamics" at work in the world.

This problem doesn't become explicit in "Signs of the Times," but it destabilizes the essay's tone. It is occasionally difficult to tell whether Carlyle is inveighing against the Age of Machinery, celebrating it, or waxing wry at its expense.

The shuttle drops from the fingers of the weaver, and falls into iron fingers that ply it faster. The sailor furls his sail, and lays down his oar; and bids a strong, unwearied servant, on vaporous wings, bear him through the waters. Men have crossed oceans by steam; the Birmingham Fire-king has visited the fabulous East; and the genius of the Cape, were there any Camoens now to sing it, has again been alarmed, and with far stranger thunders than Gamas. There is no end to machinery. Even the horse is stripped of his harness, and finds a fleet fire-horse yoked in his stead. Nay, we have an artist that hatches chickens by steam; the very brood-hen is to be superseded! . . . We remove mountains, and make seas our smooth highway; nothing can resist us. We war with rude Nature; and, by our resistless engines, come off always victorious, and loaded with spoils.[45]

The dizzying uncertainty of tone here is related to uncertainty about the location of "nature" in the passage. The last sentence tries to sum things up by saying that man, allied with mechanism, wars against Nature. But the passage began in a way that tended to suggest a rather different alignment: mechanism, allied with Nature, wars against man. The victory of mechanism is described so as to give it a quality of smooth inevitability rather than laborious artifice. "The shuttle drops from the hand of the weaver, and falls into iron fingers that ply it faster. The sailor furls his sail. . . ." As human beings resign their tasks, they are automatically—and to all appearances, naturally—taken up by machines, or even by "fire" and "steam" themselves—in the form, not of wheeled engines, but of quasi-magical "fire-horses." If Carlyle can't decide whether to celebrate this process or mourn it, this is in large part because the terms "nature" and "mechanism" are refusing to line up against each other as the Romantic argument he has inherited suggests they are supposed to do. Mechanism has come to seem natural.

In *The Machine in the Garden*, Leo Marx emphasizes Carlyle's critique of mechanism in order to contrast it against a typically American embrace of industrial technology as mediating agent between civilization and wilderness.[46] But Carlyle is not actually as hostile to industrialism as Marx makes him out to be; he shares the intuition, becoming widespread around 1829, that mechanical production is in some exciting but disturbing sense "natural." The excitement and the disturbance had less to do with American providentialism than with the disruption of class boundaries implied by a conflation of energy and toil. In *Sartor Resartus* (1833–1834), Carlyle largely abandons the opposition between mechanism and nature. The natural force of fire, the industrial force of coal and iron, and "Man's Force" are all in alignment; the emphasis falls on the productive, work-like nature of all force whatsoever. A considerable amount of uncertainty survives, however, in Carlyle's attitude toward these productivist ideas. The continuity between the force of human labor and the force of nature has double-edged implications; it becomes the theme of both rhapsody and nightmare.

In a passage that would frequently be cited by Victorian popularizers of science, Carlyle's imaginary professor Teufelsdröckh sets forth the central dogmas of productivism. He does so, incidentally, in a way that reveals the continuing importance of the sun as a synecdoche for autonomy. Chapter 8 has more to say about the survival of that Romantic symbolism in Victorian science. More important for the present chapter is the passage's insistence on class, which reveals that it aims to universalize, and extend to the working class, a naturalistic connection that was already accepted in the domain of middle-class work.

"As I rode through the Schwarzwald, I said to myself: That little fire which glows star-like across the dark-growing (*nachtende*) moor, where the sooty smith bends over his anvil, and thou hopest to replace thy lost horseshoe,—is it a detached, separated speck, cut-off from the whole Universe; or indissolubly joined to the whole? Thou fool, that smithy-fire was (primarily) kindled at the Sun; is fed by air that circulates from before Noah's Deluge, from beyond the Dogstar; therein, with Iron Force, and Coal Force, and the far stronger Force of Man, are cunning affinities and battles and victories of Force brought about: it is a little ganglion, or nervous centre, in the great vital system of Immensity. Call it, if thou wilt, an unconscious Altar, kindled on the bosom of the All; whose iron sacrifice, whose iron smoke and influence reach quite through the All; whose dingy priest, not by word, yet by brain and sinew, preaches forth the mystery of Force; nay preaches forth (exoterically enough) one little textlet from the Gospel of Freedom, the Gospel of Man's Force, commanding, and one day to be all-commanding.

"Detached, separated! I say there is no such separation: nothing hitherto was ever stranded, cast aside; but all, were it only a withered leaf, works together with all; is borne forward on the bottomless, shoreless flood of Action, and lives through perpetual metamorphoses."[47]

The fire appropriately glows like a star across the moor, since it is ultimately a portion of the sun, a portion of the Force that circulates and has always circulated through nature. But it is also a "ganglion" where this force is bound up with and wrought together with "Man's Force," an altar-fire whose "dingy priest . . . by brain and sinew, preaches forth the mystery of Force." A twofold point is being made: that the command to work is natural, and that nature, conversely, is a productive economy where "all . . . works together with all . . . and lives through perpetual metamorphoses."

But the self-hectoring tone of the passage has less to do with that argument than with the specific choice to exemplify the bond between nature and culture in the figure of a "dingy," "sooty" smith. In addressing himself as "Thou fool," Teufelsdröckh implicitly criticizes the reader's potential tendency to see a smithy-fire only as a place where "thou hopest to replace thy lost horse-shoe." He needs to insist that this fire, which could seem an emblem of servile toil, is actually an emblem of human freedom. The smith preaches that Gospel of Freedom, "not by word, yet by brain and sinew." The projection of the scene into Germany—and moreover into a remote rural area—distances the smith not just spatially but chronologically from the factories of Birmingham. But there can be little doubt that the passage centers on the sooty fire, and envisions the smith as a tender of that fire, because industrial labor is implicitly at issue. The passage already adumbrates an argument that will be made more explicitly in "Chartism" (1840): that industrial labor, however unpleasant and routinized, is dignified by association with the natural force it appropriates.

This attempt to redeem industry through association with nature is at odds, however, with another argument in the text. Uneasy about the implications of taking mechanic toil as a paradigm for human freedom, Carlyle insists on a definition of toil as a confrontation between Man and Nature, and thereby reestablishes in the productivist system a location for the autonomous will. All Force is one—and yet Iron Force and Coal Force are surpassed by "the far stronger Force of Man . . . commanding and one day to be all-commanding." The tension is subtle in this passage, but it has become an explicit internal contradiction by the time Carlyle writes the chapter on "Labour" from *Past and Present* (1843). "Work, be it never so Mammonish, mean, *is* in communication with Nature"—and yet, on the other hand, the worker must "vanquish and compel" the world, imposing order on "rough, rude, contradictory" men and things. "Nature herself is but partially for him, will be wholly against him, if he constrain her not!"[48]

These internal contradictions in productivism play a particularly central role in Carlyle's politics. By taking dingy toil as natural force and therefore as the paradigm of human freedom, Carlyle is able to idealize the actual unfreedom of the working classes; while, on the other hand, the violence demanded by confrontation between Man and Nature lends a sanction to the masterfulness of Captains of Industry. But the contradictions of productivism extend far beyond this particular paternalistic system. Nineteenth-century productivism, which tends to subsume human agency in the category of natural force (and thereby tends to compress differences of class), always generated a compensato ry triumphalism about the technological conquest of nature. The tension was active as late as 1876, for instance in the writing of Karl Marx.

Primarily, labour is a process going on between man and nature, a process in which man, through his own activity, initiates, regulates, and controls the material reactions between himself and nature. He confronts nature as one of her own forces, setting in motion arms and legs, head and hands, in order to appropriate nature's productions in a form suitable to his own wants.[49]

For Marx, human labor is "one of [nature's] own forces," and (as Rabinbach has shown) the concept of *Arbeitskraft* or "labor power" is used by both Marx and Engels to suggest that labor should be measured in units commensurate with other natural forces. Marx, then, is a productivist. But he insists equally that labor is an act of free volition through which human beings impose order on nature. Man, "through his own activity, initiates, regulates and controls" the labor process, in order to appropriate nature's productive power. If there is a contradiction here, it is one that was never satisfactorily resolved. Perhaps it would be better to speak of a tension between two moods of productivism. Productivism in the declarative mood is the mere statement or belief that labor is continuous with natural force. But this was, in practice, usually accompanied by productivism in the imperative mood: a political and ethical injunction to maximize production by appropriating an ever-greater portion of the forces of nature, and by employing those forces ever more efficiently. These two aspects of productivism coexist uneasily in *Sartor Resartus*—whose transcendental affirmation of the unity of all force goes hand-in-hand with the imperative, "Produce! Produce!" They continue to coexist, more or less uneasily, throughout the latter half of the nineteenth century.

CHAPTER 8

PRODUCTIVISM AND THE
POPULARIZATION OF THE FIRST LAW
OF THERMODYNAMICS

By the middle of the nineteenth century, science was becoming a profession, and the universe of authenticated science was growing too large for lay readers to survey it easily. Many claims important to professionals were necessarily ignored by the broader reading public. Institutions to bridge this gap took shape (magazines of "popular science," for instance). But the process of popularization was often slow, even for ideas that might seem inherently sensational. Researchers almost immediately interpreted the second law of thermodynamics, for instance, as a global sentence of death. If sources of usable energy were irreversibly dissipating as heat, the earth would become "within a finite period of time . . . unfit for the habitation of man." But although William Thomson clearly explained these chilling implications to a professional audience in 1852, the subject did not attract much public attention until the 1870s.[1] This sort of delay was not unusual. The surge of public interest in the first law of thermodynamics—known today as the conservation of energy—was by contrast rapid and intense enough to pose something of a historical puzzle.

Hermann von Helmholtz showed in 1847 that a system of masses whose motions were governed by attracting or repelling central forces would always conserve a dynamic quantity that could be expressed either as $\frac{1}{2}mv^2$ or as the integral of force across distance. As W. J. M. Rankine restated the idea in 1853, every loss of "actual energy" is a gain of "potential energy," and vice versa.[2] In less than a decade, the newly discovered principle was hailed as the foundation and capstone of all science. Herbert Spencer called it "the sole truth which transcends experience by underlying it"; for Michael Faraday it was "the highest law in physical science which our faculties permit us to perceive."[3] Articles on the topic began to appear in reviews in the late 1850s; by the early 1860s they were proliferating in popular monthlies and weeklies (*All the Year Round*, *Once a Week*, *Good Words*, and *Cornhill Magazine*, for instance) where they rubbed shoulders with serialized fiction. In an introduction to an American anthology on the topic designed for an audience of nonspecialists, Edward L. Youmans proclaimed that "the conservation of forces" was "the highest law in *all* science—the most far-reaching principle that adventuring reason has discovered in the universe."[4]

But enthusiasm for the first law coincided with deep confusion about its content. In the 1860s the new scientific doctrine was not usually called "the conservation of energy," but the "circulation," "correlation," or "persistence of force." Writers' hesitation between "energy" and "force" betrayed an uncertainty that was more than semantic. Faraday, for instance, thought the conserved quantity was the instantaneous product of mass and acceleration— a quantity that might be experienced as pressure or attraction. Faraday therefore inferred that Newton's inverse-square law could not be a complete description of gravity, since bringing two bodies nearer would seem to create, and moving them apart would seem to destroy, attractive force— and this, he thought, had been proved impossible. He used this supposed anomaly to predict a "gravelectric effect" that would be produced when the force of gravity was annulled by distance.[5] In fact, the product of mass and acceleration (ma) is not conserved; what is conserved is a quantity that can be expressed as the integral of force (ma) across distance. Spencer not only shared Faraday's mistaken idea that force is conserved but spoke about the new principle in terms that show confusion about the very concept of a conservation law—equating "the persistence of force" with the constant gravitational coefficient that has to be assumed whenever weight is used to measure mass.[6]

Researchers who were more closely connected to the professional networks working on these problems did, of course, understand that the conserved quantity was proportional to $\frac{1}{2}mv^2$ and not to ma. But even those writers often, like John Tyndall, continued to use the word "force" for both quantities until the 1870s, or made a distinction between "force" and "moving force" that was lost on a less mathematically minded audience. As a consequence, articles about "the circulation of force" continued to misinform lay readers about the conserved quantity well into the 1860s. Anne Gilchrist, for instance, agreed with Faraday that the conservation of force proves that Newton's inverse-square law is an incomplete description of gravity.[7] James Hinton defined force even more vaguely as change. "Is the quantity of force in nature (that is, of change or tendency to change) always the same? Science answers this question in the affirmative. The amount of force does not vary."[8] The word "force" was ambiguous, but a large number of Victorians were substantively, and not just semantically, confused.

The confusion is not altogether surprising. Force and power are predicates that can be experienced directly and assigned to tangible objects; force is (speaking informally) the weight one struggles to lift, power the amount of weight an agent can lift in a given time. There were niches in the English language for these concepts to fill. The integral of force across distance, by contrast, is a property of systems, and one that cannot easily be experienced directly; its practical significance as "duty" or *effet dynamique* only became clear in the late eighteenth century—and then primarily to engineers. Understandably, popularizers chose to describe the new conservation principle as if it referred to the more familiar quantity, force. But the elusive nature of the quantity eventually called energy then makes it all the more surprising that the first law was acclaimed as an epoch-making discovery. What did Victorians think had been discovered? Not, evidently, the conceptual or practical importance of $\frac{1}{2}mv^2$ in particular.

One might hypothesize that nineteenth-century readers were indiscriminately attracted to conservation principles—as emblems of the divine Providence that designed a self-sustaining world. Readers would not need to understand the conserved quantity, in that case, to appreciate the idea of conservation. David Gooding, for instance, has argued that Faraday's resistance to mechanical models of force was founded on his view of nature as a self-poised dynamic economy, which in turn was founded on his theology.[9] But although this argument is persuasive in Faraday's case, it would be difficult to generalize it to explain widespread enthusiasm for the conservation of force in the late 1850s and 1860s. Not everyone shared Faraday's Sandemanian theology, and the notion of a self-sustaining natural economy had actually become unsettling in the wake of works by Charles Lyell and Thomas Chalmers.[10] By mid-century, some Christians felt obliged to show that Nature's poise was subject to decay, and therefore dependent on occasional renovation by God. Crosbie Smith has argued that William Thomson's thermodynamic research seeks for this reason to demonstrate progression and decay in nature.[11] In short, the notion of perpetual balance was not a straightforwardly reassuring one at this historical juncture; it would not in itself have provided a reason for Presbyterians and uniformitarians, Sandemanians and agnostics to unite in agreement about the importance of energy conservation.

Productivism and the Solar Interpretation of Thermodynamics

What then explains the consensus of acclaim that rapidly coalesced around the conservation of force? I will argue that Victorians found it easy to recognize this doctrine as an exciting and fundamental physical law because it was presented as ratifying long-established habits of economic thought. In announcing the conservation of force, researchers immediately stressed that the new discovery proved that economic production and natural force were two names for a single power. This is a belief Anson Rabinbach has named "productivism": "the belief that human society and nature are linked by the primacy and identity of all productive activity, whether of laborers, of machines, or of natural forces." I am indebted to Rabinbach's exploration of this idea, and of its late-nineteenth-century social consequences. But since writers like G. G. Coriolis, Charles Babbage, J. F. W. Herschel, and Thomas Carlyle had promulgated a similar view of industry in the early 1830s, I am inclined to reject one of his historical premises: that thermodynamic science gave rise to productivism.[12] The truth is closer to the reverse. Thermodynamic science leapt to prominence so rapidly because Victorian scientists, lecturers, and journalists believed that it ratified a productivist conception of industry they already cherished.

This hypothesis does a great deal to explain Victorian writers' insistence on interpreting the first law of thermodynamics as a statement about the sun. The conservation of energy is a general dynamic principle; it has nothing to do with the sun in particular. Why was it presented to the reading public in articles with titles like "Sun Force and Earth Force," "The Labour of the Sunbeams," and "What We Owe to the Sun"? The answer, I think, is that thirty years earlier Herschel and Carlyle had traced work and natural force back to a common source in the sun—a source that grandly and visibly united natural and economic production. Victorian journalists and popular lecturers framed the first law, likewise, as a statement about the sun because they understood it, likewise, as a statement about the natural origin of labor power. Many commentators in fact found this aspect of the discovery so important that they credited the discovery of energy conservation (mistakenly) to Herschel or Carlyle or the engineer George Stephenson. The mistake is revealing: popularizers seem to have cared much more about the connection between labor and sunlight than they did about the mathematics of conservation.

To deny that thermodynamic science gave rise to productivism is not to deny thermodynamics its deservedly prominent place in Victorian cultural history. Although analogies between work and natural force had been in circulation since the late eighteenth century, researchers' ability to specify and confirm a ratio of correlation promoted those analogies to the status of a basic natural law, making them far more influential than they would otherwise have become. And although the precise mathematical definition of energy did elude most early readers, it too mattered in the long run. As a property of systems, energy was more difficult to make tangible than weight or velocity. Even writers like Faraday and Spencer failed to grasp its significance. But this very elusiveness eventually became itself a culturally significant fact. Neoclassical economists may have taken a hint from the new concept, concluding that economic value was likewise a property of systems rather than of substances—as Philip Mirowski suggests in his bold study of late-nineteenth-century economic theory.[13] More broadly, the inherent elusiveness of energy gave aid and comfort to the transcendental register of materialist thought Bruce Clarke calls "visionary scientism."[14]

But these delayed responses do little to explain the immediate enthusiasm for thermodynamic science from 1857 through 1867. "The Conservation of Force" interested readers at first because it seemed to link the natural order to the order of economic production—and especially, because it seemed to link labor to sunlight. This solar interpretation of thermodynamics began with the researchers themselves. In the *Proceedings of the Royal Society of Edinburgh* for 1852, William Thomson stressed that "Heat radiated from the sun . . . is the principal source of mechanical effect available to man."[15] His observation was widely echoed by other scientists. Hermann von Helmholtz observed, "all force, by means of which our bodies live and move, finds its source in the purest sunlight." Justus Liebig claimed that the power of the working body was "derived in the first instance from the sun."[16]

But in articles written by researchers the link between labor and the sun was usually confined to an illustrative anecdote. In more popular contexts, it set the terms of the whole discussion. "When we move machinery by steam," the *Popular Science Review* reported in 1866, "we call into requisition force which the sun has at some previous time supplied to the earth, and which we now simply liberate: indirect sun force." The same thing holds true of wind and water power, and of the labor of "the animal body."[17] The discovery proves, in the words of another writer, that "the source of all labour is the sun. All the labour done under the sun is really done by it."[18] Out of the depths of time, *Recreative Science* marveled, "the lost sunlight of an unpeopled earth lights us, warms us, and works for us with the sunshine of today."[19] "In very truth," *Once a Week* concluded, "are we 'the children of the sun.' "[20] The joy that popular-science writers found in this theme was related to its power to condense the varied forms of labor in a complex industrial society down to a radically physical, easily surveyed totality—a totality that was, indeed, self-illuminating. This is why Herbert Spencer, for instance, made "The Correlation and Equivalence of Forces" a linchpin of his social philosophy.

> Based as the life of a society is on animal and vegetal products; and dependent as these animal and vegetal products are on the light and heat of the sun; it follows that the changes going on in societies are effects of forces having a common origin with those which produce all the other orders of changes

which have been analyzed. Not only is the force expended by the horse harnessed to the plough, and by the labourer guiding it, derived from the same reservoir as is the force of the falling cataract and the roaring hurricane; but to this same reservoir are eventually traceable those subtler and more complex manifestations of force which humanity, as socially embodied, evolves.[21]

For Spencer as for many other writers, the point of correlating forces to each other lay less in the quantity conserved than in the associated implication that physical, vital, mental, and economic forces all shared a connate origin. In introducing his often-reprinted anthology on the topic, Youmans echoed Spencer's emphasis on the first law's power to unify the social and physical realms. Since "in the dynamical point of view there is a strict analogy between the individual and the social economies," even the forces of "morality and liberty" should perhaps be correlated to physical force.[22]

This fascination with the solar origin of labor had preceded thermodynamic science by more than thirty years. In *A Treatise of Astronomy* (1833)—frequently republished throughout the rest of the century as *Outlines of Astronomy*—J. F. W. Herschel digressed for a moment to make the following observation about the sun.

The sun's rays are the ultimate source of almost every motion which takes place on the earth. By its heat are produced all winds, and those disturbances in the electrical equilibrium of the atmosphere which give rise to the phenomena of terrestrial magnetism. By their vivifying action vegetables are elaborated from inorganic matter, and become, in their turn, the support of animals and of man, and the sources of those great deposits of dynamical efficiency which are laid up for human use in our coal strata. By them the waters of the sea are made to circulate in vapour through the air, and irrigate the land, producing springs and rivers.[23]

It would be an exaggeration to call Herschel's train of reasoning in this passage even incipiently thermodynamic. His concern is not to correlate quantities of heat and motion, but to locate the ultimate source of both—and especially, the source of the economically useful motion latent in wind, water, food, and coal. The science Herschel used to do this was not new; the passage synthesizes insights that were already several decades old, and that have been discussed in previous chapters of this book. In 1802 Adam Walker had argued "that all fire is originally derived from the sun," and that the burning of coal in particular released "light that has been imprisoned five thousand years."[24] French engineering theory, quite probably mediated through his friend Charles Babbage, taught Herschel how to translate "fire" into "dynamical efficiency."[25]

Herschel's insights were widely diffused in mid-Victorian literature. Thomas Carlyle stressed in *Sartor Resartus* that every smithy-fire is "(primarily) kindled at the Sun"; his investment in that image has been discussed more fully in chapter 7.[26] But the person most often cited by Victorians as the originator of this idea was the railway engineer George Stephenson (1781–1848). A number of people who knew Stephenson reported his enduring fascination with the idea that his locomotives were powered by the sun. Because this fascination is only recorded anecdotally, it is difficult to date its inception, but the statements attributed to Stephenson suggest that his thinking was shaped more by early-nineteenth-century chemistry than by thermodynamic science. Samuel Smiles reports the following conversation, which took place in 1845.

> "Now Buckland," said Mr. Stephenson, "I have a poser for you. Can you tell me what is the power that is driving that train?" "Well, said the other, I suppose it is one of your big engines." "But what drives the engine?" "Oh, very likely a canny Newcastle driver." "What do you say to the light of the sun?" "How can that be?" asked the doctor. "It is nothing else," said the engineer: "it is light bottled up in the earth for tens of thousands of years,—light, absorbed by plants and vegetables, being necessary for the condensation of carbon during the process of their growth, if it be not carbon in a different form,—and now, after being buried in the earth for long ages in fields of coal, that latent light is again brought forth and liberated, made to work, as in that locomotive, for great human purposes."[27]

This conversation was filtered through William Buckland and Samuel Smiles before appearing in print in 1857, so it is difficult to date and attribute specific turns of phrase. But the argument put in Stephenson's mouth is very credibly the kind of thing a sixty-four-year-old engineer who taught himself science as a young man might have said in 1845. Stephenson's argument—that plants imbibe light, and that coal is therefore in effect condensed or bottled light—has little relation to thermodynamics, but is entirely in keeping with the scientific theories that were in circulation at the beginning of the century.[28] In particular, Stephenson's remark that light may be a form of carbon reveals how close he remains to early-nineteenth-century photochemical thinking. This attempt to explain the phenomena of light, heat, and combustion by reference to a shared material substrate is closer to Humphry Davy than to the research on "the correlation of forces" underway in the 1840s.

Stephenson's connection between railways and sunlight seeks first of all to validate the industrial economy as a whole. Understood as sunlight, the locomotive no longer embodies the grimy self-interest of a grasping upstart—but a principle universal and, if not eternal, at least "tens of thousands

of years" old. The social logic of this analogy can be traced back to
Ricardian political economy, and—in a Satanic sense—to Wordsworth's
Excursion. What is relatively new about Stephenson's claim is its narrow
focus on motive power. Like Herschel, he revives a chain of inferences
drawn from the science and poetry of the Romantic period, but refocuses
the argument on motive power rather than a generalized principle of
enterprise, activity, or culture. In chapter 7, I explored the reasons for this
shift of emphasis, and stressed that it destabilized the boundary between
working-class and middle-class labor. Some of the less pleasant social
implications of the shift are visible in this anecdote. Stephenson's simplification of industry to sunlight in this case displaces "the canny Newcastle
driver" who might otherwise seem to be driving the engine; the real
agency of the railway concern turns out to be visible only from the perspective of an engineer.

Competing Interpretations of "The Conservation of Force"

I have argued that the first law of thermodynamics owed much of its
popularity to a perception that it ratified the unity of labor and natural force.
As evidence, I have pointed to researchers' and journalists' habit of publiciz-
ing the first law by translating it into a statement about the solar origin of
labor. Other interpretations were possible. What is interesting is that they
failed to catch on. William Grove's lectures *On the Correlation of Physical Forces*
(1843, pub. 1846) carefully resist the idea that the correlation of different
forces implies an attempt to trace force back to its ultimate origin. According
to Grove "no force can, strictly speaking, be initial, as there must be some
anterior force which produced it"[29]; Grove eschews the "search after essential causes" altogether, in favor of "a search after facts and relations."[30] His
attention is thus directed to abstract correlations, not to any single object that
might serve as a figure for natural power. Logic was in many respects on
Grove's side. But few other writers endorsed this ascetic interpretation of the
"correlation of forces," and there seems to have been little popular response
when Grove's lectures were first published in 1846.[31] Later, especially in the
period between 1857 and 1867, there would be a positive deluge of articles
on the topic. But what evoked fascination at that point was precisely the
question Grove had eschewed: what is the fountain and ultimate source of all
terrestrial force?

The first lengthy discussion of Grove's work to appear in a popular
British periodical was "The Phasis of Force" (1857), by William Benjamin
Carpenter, a physiologist. His discussion in some respects marks a transitional

stage between Grove's own interpretation of the "correlation of forces" and the sun-centered interpretation that caught the popular imagination around 1862. For Carpenter, the importance of the correlation of different forces was that it underlined a "phasis," or perpetual mutual transformation, between the forces of the inanimate world and such phenomena as "animal heat," "Muscular contraction," and "Nerve-force." But Carpenter's primary example of this dynamic continuity is the transformation of sunlight into chemical potential by plants, and its reproduction as warmth both in animal bodies and coal-fires.

> [T]he warm-blooded animal may be said to be continually restoring to the universe that force which the growing plant had appropriated to itself. And carrying the same principle a little further, we may say that in utilising the stores of coal which have been prepared by the luxuriant vegetation of past ages . . . man is actually reproducing and applying to his own uses the light and heat which its vegetation drew from the solar beams. . . .[32]

Although Carpenter's term "phasis" purports to define a perpetual circulation of force, his discussion in practice begins with the sun and follows a series of transformations downward, implicitly referring agency upward to the sun itself. Carpenter credits George Stephenson as the first person to have grasped the sun's importance. "Looking at this matter from the standpoint afforded by the 'correlation' doctrine . . . we cannot but feel an increased admiration of the intuitive sagacity of that remarkable man George Stephenson, who was often laughed at for propounding in a somewhat crude form the very idea which we have just been endeavouring to present under a more philosophical aspect."[33]

This nod to Stephenson is typical of early writing on the "conservation of force."[34] It isn't necessarily a sign that Stephenson was in any direct sense an influence on Carpenter. Stephenson had acquired folk-hero status in Britain by the time of his death in 1848, and Samuel Smiles' immensely popular biography (which went through five editions in 1857, the first year of publication) helped to reinforce that status. Victorians had a high regard for knowledge gained through practical experience—and a high regard for railways in particular. Many writers were happy to attribute an important insight about coal to this railway engineer, and liked to imply that he had stumbled upon it in the course of his eminently practical labors. (He is more likely to have borrowed it from Herschel or from Romantic-era chemistry.) For a while, therefore, many people who wrote about the "conservation of force" in Britain said that this new doctrine was first grasped when George Stephenson realized that his locomotives were driven by bottled sunlight. This shows that Victorian writers interpreted the first law above all as a statement about the connection between nature and industrial power, but it need not actually imply that they were borrowing that interpretation from Stephenson. Innumerable other sources (among them Davy, Byron, Carlyle, Herschel) had encouraged Britons to use the sun as a sign of the unity of all terrestrial power.

The attention given to Stephenson's "intuitive sagacity" understandably rattled John Herschel in particular, who had remarked in 1833 that "the sun's rays are the ultimate source of almost every motion which takes place on the earth."[35] By 1851, Herschel had already begun to suspect that his passing remark of 1833 had acquired a new importance; in reprinting his *Outlines of Astronomy* that year, he underlined his claim to chronological priority by adding a footnote—"So in the edition of 1833"—to the relevant paragraph about the sun, after the sentence about the "dynamical efficiency" of coal.[36] In editions published after 1857 a further sentence is added to the footnote: "The celebrated engineer Stephenson is generally, but erroneously, cited as the originator of this remark."[37] The message appears to have gotten through to Samuel Smiles, because he eventually deleted a sentence that, in the 1857 edition of his biography, had described Stephenson's idea as "a most striking and original one."[38] Herschel's growing possessiveness about his remark of 1833 reveals a growing awareness that it had become something worth fighting over. Herschel's productivism, which was nothing more than a passing observation in 1833, had gained additional prestige from new thermodynamic research. But the prestige ascribed to the remark also reveals how many Victorians interpreted the conservation of energy as a claim about the connection between labor and the sun.

John Tyndall echoed Herschel's remarks in the following passage, delivered in 1862 in a series of lectures at the Royal Institution, and later reprinted as *Heat as a Mode of Motion* (1863).

> Leaving out of account the eruptions of volcanoes, and the ebb and flow of the tides, every mechanical action on the earth's surface, every manifestation of power, organic and inorganic, vital and physical, is produced by the sun. His warmth keeps the sea liquid, and the atmosphere a gas, and all the storms which agitate both are blown by the mechanical force of the sun. He lifts the rivers and the glaciers up to the mountains; and thus the cataract and the avalanche shoot with an energy derived immediately from him.[39]

Tyndall acknowledges a debt to Herschel—writing in a footnote that "The germ, and much more than the germ, of what is here stated is to be found in a paragraph in Sir John Herschel's *Outlines of Astronomy*."[40] Tyndall echoes

Herschel not only in his central claim, but in his use of anaphora (repetition of the beginning of a sentence) to emphasize the sun's ubiquity. There are important rhetorical changes; the most important is Tyndall's freer use of personification. Tyndall makes the personified (and masculine) sun the subject of each clause—"He lifts the rivers and the glaciers . . . "—whereas Herschel had written in the passive voice about a neuter sun, beginning his sentences with phrases like "By its heat . . . By their vivifying action . . ." Much of the reason for this rhetorical difference lies in Tyndall's aggressive naturalism.[41] Tyndall's personification of the sun erects a natural deity to replace the Christian God; Herschel, a believing Anglican, would not consciously have sought to imply this.

Tyndall goes on to use the same strategy of personification in order to represent the sun as the agent of all economic production. The germ of this idea, too, is present in Herschel's description of the sun as the source of the "dynamical efficiency" stored up in coal, but Tyndall develops it with greater verve and at greater length.

> The clover sprouts and blossoms, and the scythe of the mower swings, by the operation of the same force. The sun digs the ore from our mines, he rolls the iron; he rivets the plates, he boils the water; he draws the train. He not only grows the cotton, but he spins the fibre and weaves the web. There is not a hammer raised, a wheel turned, or a shuttle thrown, that is not raised, and turned, and thrown by the sun.[42]

Again there are echoes of Christian rhetoric here, rising on occasion to such a pitch that pious readers must have suspected blasphemy—for instance, when Tyndall remarks of the sun (on the same page) that "the lilies of the field are his workmanship. . . ." Tyndall had been brought up in a strict Protestant family before he underwent a conversion to agnostic naturalism. The wry pleasure he takes in skirting blasphemy is undoubtedly mixing here with another sort of pleasure, springing from the familiar sermonic cadences themselves.

Tyndall's solar interpretation of the world takes up, in all, the last four and a half pages of *Heat as a Mode of Motion*. It is the most eloquent and longest such passage in Victorian literature, and it may well have been instrumental in making a productivist interpretation of the sun a central part of late-Victorian thinking. Though, like many influential popularizations of science, the book has slipped through the disciplinary grid of historical study, Tyndall's *Heat* was important in its own time. Writing in 1912, J. T. Merz remarked, "no publication did more to establish a general kinetic view of matter and of natural phenomena. . . . The book was to the popular mind a revelation; it was translated into many foreign languages, ran through many editions, was recommended by thinkers of the first order, and the title coveted as 'manifesting far and wide through the world one of the greatest discoveries of modern philosophy.' "[43] Tyndall's encomium on the sun, in particular, seems to have made a considerable impression. In America, Henry Adams called it a "famous outburst of eloquence."[44] The Victorian philologist and mythographer Max Müller used Tyndall's book to back up his argument that religion began as a personification of the sky and of the sun. Tyndall has shown, Müller writes, "how we live, and move, and have our being in the sun, how we burn it, how we breathe it, how we feed on it . . ."[45]

In short, the popular response to the conservation of force in Britain suggests that Victorians were at first less impressed by the—poorly elucidated—concept of energy than by the claim that "all the labour done under the sun is really done by it."[46] A priority dispute between the advocates of George Stephenson and John Herschel tends to indicate that the discovery of a connection between sunlight and industry was held to be as important as the conservation of force itself. In fact, the distinction between the conservation principle and the solar productivism that preceded it was usually blurred, even by a writer like Tyndall, who could have explained the distinction. In "Personal Recollections of Thomas Carlyle" (1890), Tyndall actually gives Carlyle credit for anticipating the conservation of energy.[47] Tyndall quotes the "Schwarzwald" passage from *Sartor Resartus* at length. After Carlyle's words "Thou fool, that smithy-fire was primarily kindled at the sun," he inserts, in parentheses, "Joule and Mayer were scientifically unborn when these words were written." At the end of the quotation he remarks that "such passages—and they abound in his writings—might justify us in giving Carlyle the credit of poetically, but accurately, foreshadowing the doctrine of the Conservation of Energy."[48] As a remark on the history of conservation ideas (poetical or not) this is misleading. Carlyle said nothing in the "Schwarzwald" passage about quantities of force; he was arguing for a productivist interpretation of nature, not for a conservation principle. But in giving too much credit to Carlyle, Tyndall inadvertently reveals that the primary significance of the conservation of energy—as far as he was concerned—lay in the productivist rhetoric it ratified, and especially in the fact that it referred all power and production back to the single fountain of the sun.

In this book I have focused on the history of productivism before 1860, because I feel that part of the story is not yet well understood. There are by contrast several excellent studies of productivism in the latter half of the nineteenth century, and it seems best to end by referring the reader to those sequels. Anson Rabinbach (who in many respects opened up this area of inquiry) has argued that analogies between economic and natural production fostered sciences of efficiency and fatigue that reshaped the working day for millions of laborers.[49] Bruce Clarke and Mark Seltzer have shown how similar analogies complicated subjectivity in late nineteenth-century fiction, and fostered a new confidence in the congruence of physical and social agency.[50]

What about the fate of productivism in the twentieth and twenty-first centuries? The title of a recent collection edited by Bruce Clarke, *From Energy to Information*, aptly sums up one important direction of change. As the economies of older industrialized nations restructure to emphasize their service sectors, and a professional-managerial class consolidates its collective identity, the concept of information has acquired a social prominence that tends to subsume the functions of older analogies between work and physical energy. The economic realm is now conceived less as a circulatory extension of nature than as a virtual image of it—representing physical existence, controlling it, and perhaps (it is feared) replacing it. By linking energy to information, disciplines like cybernetics and statistical mechanics facilitate this sublation of the physical world—or do so, at least, in theory.[51]

In practice, of course, several different models of social life coexist and compete for attention at any given time. An oil embargo gave the concept of energy political urgency in the 1970s, and fostered a distinct revival of productivist rhetoric. Not surprisingly, the sun became once again an emblem of personal and national autonomy—partly through new fantasies about self-sufficient solar houses, and partly (as in the nineteenth century) because writers used dynamic abstractions to symbolically ground social life in nature. I have before me a book about the sun published by the Smithsonian Institution in 1981 that reads in many places as though it had been written by John Tyndall.

> All of observed life is bound up in the fact of the existence of the sun, our "onlie begeter." Everything that happens, from rainbows to the fall of sparrows, from snow to the winds that let us sail or fly, from the appearance of civilizations to their demise in the sands of time—all is dependent on our star. The paper on which this book is printed, the light by which it is read, and the hand which writes these words are all evidence of the "fire of life."[52]

And yet this passage is not quite Tyndall. In place of his systematic emphasis on industrial hammers, wheels, and shuttles, it has a network of literary and historical allusions—to *Hamlet* and "This Living Hand," "Ozymandias" and the "onlie begetter" of Shakespeare's sonnets. I think the insistent literariness of the passage is a clue that these ideas have come to know their own age. Twentieth- and twenty-first-century writers still naturalize social life by translating it into sunlight. But the gesture does not carry the same weight it once carried. In an era when the most notorious fortunes are made not off locomotives or oil but off software and energy futures, this insistence on the primacy of the physical world evokes an unstated nostalgia.

NOTES

Introduction

NOTES TO PP. 1–4

1. "The Source of Labour," *Chambers's Journal*, 4th series, 3 (1866): 555–56. John Tyndall, *Heat Considered as a Mode of Motion* (New York: D. Appleton, 1864), 446–47.

2. Anson Rabinbach, *The Human Motor: Energy, Fatigue, and the Origins of Modernity* (New York: Basic Books, 1990), 3, 69–83, 179–237. Karl Marx, *Capital: A Critique of Political Economy*, trans. Ben Fowkes (New York: Vintage, 1977), 283. For the late-nineteenth-century literary ramifications of productivism, see Mark Seltzer, *Bodies and Machines* (New York: Routledge, 1992); also Bruce Clarke, *Energy Forms: Allegory and Science in the Era of Classical Thermodynamics* (Ann Arbor: University of Michigan Press, 2001).

3. Rabinbach, *The Human Motor*, 3.

4. John Milton, *Paradise Lost*, ed. Scott Elledge (New York: W. W. Norton, 1975), 94 (4:618–22).

5. William Godwin, *Caleb Williams*, ed. David McCracken (London: Oxford University Press, 1970), 218–19.

6. Isaac Newton, *Opticks* (New York: Dover, 1952), 401. See also P. M. Heimann and J. E. McGuire, "Newtonian Forces and Lockean Powers: Concepts of Matter in Eighteenth-Century Thought," *Historical Studies in the Physical Sciences* 3 (1971): 233–306.

7. Stephen Wallech, " 'Class Versus Rank': The Transformation of Eighteenth-Century English Social Terms and Theories of Production," *Journal of the History of Ideas* 47 (1986): 409–31. Isaac Kramnick, *Republicanism and Bourgeois Radicalism: Political Ideology in Late Eighteenth-Century England and America* (Ithaca: Cornell University Press, 1990), 60. The phrase "republican tradition" is appropriate in this context, although the assumptions under discussion could in principle be adapted to non-republican forms of government. See John Robertson, "The Scottish Enlightenment at the Limits of the Civic Tradition," *Wealth and Virtue: The Shaping of Political Economy in the Scottish Enlightenment*, ed. Istvan Hont and Michael Ignatieff (Cambridge: Cambridge University Press, 1983), 138.

8. William Hazlitt, *The Complete Works*, ed. P. P. Howe, 21 vols. (London: J. M. Dent, 1930), 5:3, 5:1.

9. Jerome McGann, *The Poetics of Sensibility* (Oxford: Oxford University Press, 1996). Geoffrey Hartman, *Wordsworth's Poetry 1787–1814* (Cambridge, MA: Harvard University Press, 1987).

10. David Simpson, from whom I draw the phrase "agrarian idealism," has described a very similar tension. See Simpson, *Wordsworth's Historical Imagination: The Poetry of Displacement* (New York: Methuen, 1987), 56–78.

11. Adam Smith, *An Inquiry into the Nature and Causes of the Wealth of Nations*, ed. R. H. Campbell, A. S. Skinner, and W. B. Todd, 2 vols. (Oxford: Oxford University Press, 1976), 1:364–65.

12. Raymond Williams, *The Country and the City* (New York: Oxford University Press, 1975), 127, 130–31.

13. Leo Marx, *The Machine in the Garden: Technology and the Pastoral Ideal in America* (1964; Oxford: Oxford University Press, 2000), 227–42, 271n, 268–319.

14. Tyndall, *Heat*, 447.

15. *PW* 2:142–44. David Ricardo, *On the Principles of Political Economy and Taxation, Works and Correspondence*, ed. Pierro Sraffa and M. H. Dobb, 10 vols. (Cambridge: Cambridge University Press, 1966), 1:69.

16. Antoine Lavoisier and Armand Seguin, "Premier mémoire sur la respiration des animaux," in Antoine Lavoisier, *Œuvres*, 7 vols. (Paris: Imprimerie impériale, 1862), 2:688–703.

17. G. W. F. Hegel, *The Philosophy of History*, trans. J. Sibree (New York: Dover, 1956), 103.

18. Annabel Patterson, "Wordsworth's Georgic: Genre and Structure in *The Excursion*," *The Wordsworth Circle* 9 (1978): 145–54.

19. Simpson, *Wordsworth's Historical Imagination*, 56–78.

20. John F. W. Herschel, *A Treatise of Astronomy* (London: Longman, 1833), 211. Thomas Carlyle, *The Works of Thomas Carlyle*, 30 vols. (New York: Scribner's, 1896), 1:56.

Chapter 1   Light as Fluid Agency

1. Kepler thought the sun sustained the motion of surrounding worlds through its own rotation. See Johannes Kepler, *New Astronomy*, trans. William H. Donahue (Cambridge: Cambridge University Press, 1992), 379–81. Writers in the first half of the eighteenth century drew a different analogy to the human heart by suggesting that the sun's emission of light and absorption of air produced a circulatory current. See, for instance, Richard Lovett, *Philosophical Essays* (Worcester: R. Lewis, 1766), 233–39.

2. Isaac Newton, *Opticks* (New York: Dover, 1952), 401.

3. Michel Foucault, *The Order of Things: An Archaeology of the Human Sciences* (New York: Random House, 1970).

4. Stephen Wallech, " 'Class Versus Rank': The Transformation of Eighteenth-Century English Social Terms and Theories of Production," *Journal of the History of Ideas* 47 (1986): 409–31.

5. Versions of this phrase are scattered widely in eighteenth-century literature. This one happens to come from James Thomson, "Liberty," *The Complete Poetical Works of James Thomson,* ed. J. Logie Robertson (London: Oxford University Press, 1908), 409 (5:604).

6. This list is not exclusive. Approaching the question from the perspective of natural history, one could just as reasonably emphasize the disappearance of an informing vegetative principle that was distinct from common matter on the one hand and from the soul on the other. Mid-eighteenth-century debates over living form dispense with this hypothesis, preferring to combine something preexisting (whether a germ or an "internal mold") with attractive forces that set it in motion. See James L. Larson, *Interpreting Nature: The Science of Living Form from Linnaeus to Kant* (Baltimore: Johns Hopkins University Press, 1994), 132–69.

7. See e.g., Jean Baptiste van Helmont, "Spiritus vitae," *Ortus Medicinae* (Amsterdam, 1648), 195–201. The role of solar symbolism in alchemy is discussed in B. J. T. Dobbs, *The Janus Faces of Genius: The Role of Alchemy in Newton's Thought* (Cambridge: Cambridge University Press, 1991), 38–45.

8. Hermann Boerhaave, *Elements of Chemistry, Being the Annual Lectures of Herman Boerhaave, M. D.* (London: J. & J. Pemberton, 1735), 1:47; see also 1:37. Boerhaave taught medicine and chemistry at the University of Leyden; his students went on to teach at Scottish and English universities, often using Boerhaave's lectures as a text.

9. Isaac Newton, "Of Natures Obvious Laws & Processes in Vegetation," *The Janus Faces of Genius: The Role of Alchemy in Newton's Thought,* by B. J. T. Dobbs (Cambridge: Cambridge University Press, 1991), 265.

10. Newton, "Of Natures Obvious Laws & Processes in Vegetation," 267.

11. Newton, *Opticks,* 374. Stephen Hales, *Vegetable Staticks* (1727; rpt. London: Oldbourne, 1961), 186–87.

12. Newton, *Opticks,* 401.

13. Newton, *Opticks,* 374.

14. For a fuller discussion of the changing meaning of "active principles" in Newton's career see J. E. McGuire, "Force, Active Principles, and Newton's Invisible Realm," *Ambix* 15 (1968): 154–208.

15. Newton, *Opticks,* 348–54.

16. P. M. Heimann, " 'Nature is a Perpetual Worker': Newton's Æther and Eighteenth-Century Natural Philosophy," *Ambix* 20 (1973): 4–7.

17. Herman Boerhaave (1668–1738) formed a theory of fire that was indebted to Newton's *Opticks,* and J. T. Desaguliers's *Course of Experimental Philosophy* (1734–1744) represented Newton's queries from the *Opticks* as if they were confirmed truth. See I. Bernard Cohen, *Franklin and Newton, Memoirs of the American Philosophical Society,* vol. 43 (Philadelphia: American Philosophical Society, 1956), 232, 254.

18. J. T. Desaguliers, *A Course of Experimental Philosophy,* 2 vols. (London: J. Senex, 1734), 2:36.

19. Hales, *Vegetable Staticks,* 178.

20. Gabrielle Emilie Le Tonnelier de Breteuil, marquise du Châtelet, *Dissertation sur la nature et la propagation du feu* (Paris: Prault, 1744), 39.

21. Robert Schofield, *Mechanism and Materialism: British Natural Philosophy in an Age of Reason* (Princeton: Princeton University Press, 1970), 157–90.

22. Du Châtelet, 39.

23. Gaston Bachelard, *The Psychoanalysis of Fire*, trans. Alan C. M. Ross (Boston: Beacon Press, 1964).

24. Bachelard, *The Psychoanalysis of Fire*, 40.

25. John Locke, *An Essay Concerning Human Understanding*, ed. A. C. Fraser, 2 vols. (New York: Dover, 1959), 1:166–82 (book. 2, chap. 8).

26. The history of this argument is explored in P. M. Heimann and J. E. McGuire, "Newtonian Forces and Lockean Powers: Concepts of Matter in Eighteenth-Century Thought," *Historical Studies in the Physical Sciences* 3 (1971): 233–306.

27. Cadwallader Colden, *The Principles of Action in Matter* (London: R. Dodsley, 1751), 16, 2.

28. One particularly influential example was Joseph Priestley, *Disquisitions Relating to Matter and Spirit* (London: J. Johnson, 1777). Priestley's matter theory, drawn from Roger Joseph Boscovich, is very different from Colden's. But he likewise deploys empirical epistemology to show that observers never encounter a strictly passive substance. Knowledge of matter or substance must therefore be knowledge of active power.

29. James Hutton, *Dissertations on Different Subjects in Natural Philosophy* (Edinburgh: Strahan, 1792), 280.

30. Hutton, *Dissertations*, 412.

31. I use "skeptical empiricism" here and elsewhere as shorthand for "a skeptical application of empiricist epistemology." These arguments are "skeptical" not in the sense that they finally deny the possibility of knowledge, but because their operative strategy is to ask whether an observer can really "form any conception" of something (say, solidity); if not, it is discarded as a superfluous hypothesis. They are "empiricist" because their criteria of knowledge always begin (and often end) in sense experience.

32. Both of the chemists discussed here—Priestley and Senebier—were in fact skeptical empiricists in matters of epistemology. I have no evidence that their epistemology determined their program of pneumatic research, though it certainly governed Priestley's presentation of that research.

33. Joseph Priestley, *Experiments and Observations on Different Kinds of Air*, 3 vols. (London: J. Johnson, 1774–1777), 1:277–78.

34. Jan Ingenhousz, *Experiments Upon Vegetables* (London: P. Elmsley, 1779), 28–30.

35. Jean Senebier, *Mémoires physico-chymiques sur l'influence de la lumière solaire pour modifier les êtres des trois règnes de la nature, et sur-tout ceux du règne végétal*, 3 vols. (Geneve: Chiral, 1782), 1:376–80. For a fuller discussion of Senebier's chemical mechanisms, see Leonard K. Nash, *Plants and the Atmosphere* (Cambridge: Harvard University Press, 1952), 71–95.

36. Senebier, *Mémoires*, 1:viii.

37. Since this chapter has suggested that skeptical empiricism fostered the notion that light might unite sensation and agency, it may be worth noting that Senebier wrote a three-volume study of "the art of observation and of making experiments." This work begins with the empirical premise that "the senses are the only bonds linking man to the objects that surround him"; it emphasizes a skeptical interpretation of Newton's first rule of reasoning; and it describes vision as the most valuable of all the senses for the observer. Jean Senebier, *Essai sur l'art d'observer et de faire des expériences*, 2nd ed., 3 vols. (Geneva: 1802), 1:1, 1:97–100, 1:188, 2:98.

38. Senebier, *Mémoires*, 3:1–239, 2:213–14. See 2:185 for Senebier's attempt to explain why light creates color in living tissue but bleaches and destroys color once that tissue is dry.

39. Senebier, *Mémoires*, 2:244.

40. Jean Senebier, *Physiologie végétale, contenant une description des organes des plantes* (Geneva: 1799–1800), 1:4–5, 2:307.

41. Antoine-Laurent Lavoisier, *Elements of Chemistry*, trans. Robert Kerr (1790; rpt. New York: Dover, 1965), 183–84.

42. Thomas Beddoes, "Observations on the Character and Writings of John Brown, M. D.," *The Elements of Medicine*, by John Brown, ed. Thomas Beddoes, 2 vols. (London: J. Johnson, 1795), 1:cxxix–cxxxiv.

43. John Brown, *The Elements of Medicine*, ed. Thomas Beddoes, 2 vols. (London: J. Johnson, 1795), 1:140.

44. Guenter B. Risse, "Brunonian Therapeutics: New Wine in Old Bottles?" *Brunonianism in Britain and Europe*, ed. W. F. Bynum and Roy Porter, *Medical History* (London: Wellcome Institute, 1988), 46–62.

45. Christoph Girtanner, "Sur l'Irritabilité, considerée comme principe de vie dans la Nature organisée," *Observations sur la physique* 36 (1790): 422–40 and 37 (1790): 139–54.

46. Thomas Beddoes, *Observations on the Nature and Cure of Calculus, Sea Scurvy, Consumption, Catarrh, and Fever: Together with Conjectures Upon Several Other Subjects of Physiology and Pathology* (Philadelphia: T. Dobson, 1797), 45, 49–50, 106.

47. Claude Louis Berthollet, "De l'influence de la lumière," *Journal de physique, de chimie, d'histoire naturelle et des arts* 29 (1786): 84.

48. Antoine-Laurent Lavoisier and Armand Seguin, "Premier mémoire sur la respiration des animaux," *Œuvres*, by Antoine-Laurent Lavoisier, vol. 2 (Paris: Imprimerie impériabe, 1862), 2:692.

49. John Thelwall, *An Essay Towards a Definition of Animal Vitality* (London, 1793), 39–40. Nicholas Roe, " 'Atmospheric Air Itself': Medical Science, Politics, and Poetry in Thelwall, Coleridge, and Wordsworth," *1798: The Year of the Lyrical Ballads*, ed. Richard Cronin (Houndmills: Macmillan, 1998), 185–202.

50. Benjamin Waterhouse, *On the Principle of Vitality* (Boston: Fleet, 1790).

51. Joseph Trent, *An Inquiry Into the Effects of Light in Respiration* (Philadelphia: Way and Graff, 1800). T. Gale, *Electricity, or Ethereal Fire* (Troy: Moffitt & Lyon, 1802), 15, 89.

52. The correlation Lavoisier and Seguin describe is complex. The number of the experimental subject's heartbeats (considered not as a rate, but as a total count) is proportional to the product of weight and height. The consumption of oxygen, in turn, is proportional to the product of the number of heartbeats and the number of breaths taken.
53. Lavoisier and Seguin, "Premier Memoire," 2:697.
54. Jean-Pierre Poirier, *Lavoisier: Chemist, Biologist, Economist*, trans. Rebecca Balinski (Philadelphia: University of Pennsylvania Press, 1996), 170–73.
55. Bernadette Bensaude-Vincent, *Lavoisier: Mémoires d'une révolution* (Paris: Flammarion, 1993), 215.

Chapter 2   *Energy* and the Autonomy of Middle-Class Work

1. Ted Underwood, "Productivism and the Vogue for 'Energy' in Late Eighteenth-Century Britain," *Studies in Romanticism* 34 (1995): 106–08.
2. The Project for American and French Research on the Treasury of the French Language, University of Chicago, World-Wide Web <http://humanities.uchicago.edu/ARTFL/ARTFL.html/> October 1995.
3. British parodies of energy-talk are cited later in this chapter. For French examples, see Michel Delon, *L'idée d'énergie au tournant des Lumières (1770–1820)* (Paris: Presses Universitaires de France, 1988), 143–49. In this otherwise indispensable account, Delon denies that there was a connection between *énergie* and middle-class aspiration, remarking that energy cannot have been a middle-class value because it "is capable of opening up expenditure and excess, and of subverting any principle of order, of wise management, of regulated production" (33). This objection is based on a caricature of nineteenth-century middle-class values. It was not "regulated production," but individual autonomy, that middle-class writers idealized in the late eighteenth century.
4. For more on the connection between energy and spontaneity, see Stuart Peterfreund, "The Re-Emergence of Energy in the Discourse of Literature and Science," *Annals of Scholarship: Metastudies of the Humanities and Social Sciences* 4 (1986–1987): 22–53.
5. Steven Shapin, "Social Uses of Science," *The Ferment of Knowledge: Studies in the Historiography of Eighteenth-Century Science*, ed. G. S. Rousseau and Roy Porter (Cambridge: Cambridge University Press, 1980), 134–39. See also Simon Schaffer, "States of Mind: Enlightenment and Natural Philosophy," *The Languages of Psyche*, ed. G. S. Rousseau (Berkeley and Los Angeles: University of California Press, 1990), 233–90.
6. Aristotle, *Metaphysics*, Loeb Classical Library, 2 vols. (Cambridge: Harvard University Press, 1936), 11.9.
7. Aristotle, *The Art of Rhetoric*, Loeb Classical Library (Cambridge: Harvard University Press, 1939), 3.11.
8. For a fuller discussion, see Delon, *L'idée d'énergie*, 39–45.

9. Sir Philip Sidney, "Defence of Poesie," *Complete Works*, ed. A. Feuillerat, vol. 3 (Cambridge: Cambridge University Press, 1923), 41.

10. Henry Fielding, *The Covent-Garden Journal and a Plan of the Universal Register-Office* (Middletown: Wesleyan University Press, 1988), 185. I am indebted to Martin Price's discussion of the connection between energy and sentimental moral philosophy in Fielding. Price persuasively connects this to a discussion of Fielding's narrative forms. Martin Price, *To the Palace of Wisdom: Studies in Order and Energy from Dryden to Blake* (New York: Doubleday, 1964), 288.

11. Henry Fielding, *The History of Tom Jones, a Foundling*, ed. Martin C. Battestin and Fredson Bowers, 2 vols. (Oxford: Weslayan University Press, 1975), 2:687 (book 13, chap. 1).

12. Denis Diderot, "Essai sur le mérite et le vertu," *Œuvres Complètes*, vol. 1 (Paris: Hermann, 1975), 100.

13. Jean-Jacques Rousseau, *La nouvelle Héloïse*, vol. 2 (Paris: Hachette, 1925), 273.

14. Ann Radcliffe, *The Mysteries of Udolpho*, ed. Bonamy Dobrée (New York: Oxford University Press, 1992), 182.

15. Radcliffe, *The Mysteries*, 329.

16. See John W. Yolton, *Thinking Matter: Materialism in Eighteenth-Century Britain* (Minneapolis: University of Minessota Press, 1983).

17. John Locke, *An Essay Concerning Human Understanding*, ed. A. C. Fraser, 2 vols. (New York: Dover, 1959), 1:312.

18. In the Boyle lecture for 1698, John Harris remarks, "*Body* or *Matter* is a Sluggish, Insensible, Passive, and Unintelligent Thing, not possibly able to move of itself, or to perform any thing by its own Power. . . ." John Harris, *The Atheist's Objections, against the Immaterial Nature of God, and Incorporeal Substances, Refuted* (London: 1698), 5.

19. J. E. McGuire, "Force, Active Principles, and Newton's Invisible Realm," *Ambix* 15 (1968): 154–208.

20. James Hutton, *A Dissertation Upon the Philosophy of Light, Heat, and Fire* (Edinburgh: Cadell, 1794), 286.

21. Robert Young, *An Essay on the Powers and Mechanism of Nature* (London: T. Becket, 1788), xi.

22. James MacLurg, *Experiments Upon the Human Bile* (London: T. Cadell, 1772), xliv.

23. The Leyden jar is a simple condenser, with two surfaces bearing opposite charges separated by glass. For its effect on electrical demonstrations and electrical theory, see J. L. Heilbron, *Electricity in the Seventeenth and Eighteenth Centuries* (Berkeley: University of California Press, 1979), 313–16. For mid-century testimony to the importance of these demonstrations, see Richard Lovett, *Philosophical Essays* (London: J. Johnson, 1766), v, 7, 240.

24. Alexander Wilson, "An Answer to the Objections Stated by M. De La Lande," *Philosophical Transactions* 73 (1783): 160.

25. Radcliffe, *The Mysteries*, 329.

26. William Hayley, *Poems and Plays by William Hayley*, vol. 3 (London: T. Cadell, 1788), 81, 96.

27. Edmund Burke, "A Letter to John Farr and John Harris, Esqrs., Sheriffs of the City of Bristol, on the Affairs of America," *The Works of the Right Honorable Edmund Burke*, 12 vols. (London: George Bell & Sons, 1875–1876), 2:31.

28. The rest of this chapter focuses strictly on the connotations of *energy* in Britain. In France the political situation was rather different; as early as 1779 *énergie* is taken as a verbal symptom of unbridled "enthousiasme." See, for instance, an interchange of letters between Madame du Deffand and the Duchesse de Choiseul discussed in Anne-Marie Jaton, "Énergétique et féminité (1770–1820)," *Romantisme* 46 (1984): 15–25.

29. Humphry Davy, notebook 20a, ms., Humphry Davy Papers, Royal Institution of Great Britain, London, 27.

30. Davy, notebook 20a, 26.

31. Isaac Kramnick, *Republicanism and Bourgeois Radicalism: Political Ideology in Late Eighteenth-Century England and America* (Ithaca: Cornell University Press, 1990), 60.

32. Kramnick, *Republicanism,* 62–63.

33. J. G. A. Pocock, *Virtue, Commerce, and History* (Cambridge: Cambridge University Press, 1985), 260–61.

34. Kramnick, *Republicanism,* 60.

35. Middle-class identity was also invoked by other groups, for other political ends. For an exploration of the widely-diverging interests served by the "middle-class idiom" in the 1790s, see Dror Wahrman, *Imagining the Middle Class: The Political Representation of Class in Britain, c. 1780–1840* (Cambridge: Cambridge University Press, 1995), 31–73.

36. J. G. A. Pocock, *The Machiavellian Moment: Florentine Political Thought and the Atlantic Republican Tradition* (Princeton, N.J.: Princeton University Press, 1975), 391.

37. Pocock, *Machiavellian Moment,* 427–32.

38. Pocock, *Machiavellian Moment,* 445–48.

39. Stephen Wallech, " 'Class Versus Rank': The Transformation of Eighteenth-Century English Social Terms and Theories of Production," *Journal of the History of Ideas* 47 (1986): 409–31.

40. Henry Brooke, *The Fool of Quality; or, the History of Henry Earl of Moreland,* 2nd ed., 5 vols. (London: W. Johnston, 1767–70), 5:36–37. See also the discussion of Brooke in Markman Ellis, *The Politics of Sensibility: Race, Gender and Commerce in the Sentimental Novel* (Cambridge: Cambridge University Press, 1996), 129–59.

41. H. T. Dickinson, *Liberty and Property: Political Ideology in Eighteenth-Century Britain* (New York: Holmes and Meier, 1977), 226–27. "No taxation without representation" was a slogan with almost as much resonance in England as in the colonies; see, for instance, John Cartwright, *Take Your Choice!* (London: J. Almon, 1776), 20–21.

42. This new model of the political significance of work also generated newly intense anxieties about idleness. See Sarah Jordan, *The Anxieties of Idleness: Idleness in Eighteenth-Century British Literature and Culture* (Lewisburg: Bucknell University Press, 2003).

43. Thomas Holcroft, *The Adventures of Hugh Trevor*, ed. Seamus Deane (London: Oxford University Press, 1973), 3.

44. Holcroft, *The Adventures*, 122.

45. Holcroft, *The Adventures*, 97.

46. William Godwin, *Caleb Williams*, ed. David McCracken (London: Oxford University Press, 1970), 218–19.

47. Paul Keen, *The Crisis of Literature in the 1790s: Print Culture and the Public Sphere* (Cambridge: Cambridge University Press, 1999), 98, 81–99.

48. Magali Sarfatti Larson, *The Rise of Professionalism: A Sociological Analysis* (Berkeley: University of California Press, 1977), 1–5. For an argument challenging this conclusion—though, I think, unconvincingly—see Wilfrid Prest, "The Professions and Society in Early Modern England," *The Professions in Early Modern England*, ed. Wilfrid Prest (London: Croom Helm, 1987), 1–24. Prest establishes that the eighteenth-century professions often recruited their members from the middle classes, and also that they served the middle classes. But this evidence does not address the decisive question: whether the market for professional services was stratified according to an autonomous logic, or according to the heteronomous principle of gentility.

49. W. J. Reader, *Professional Men: The Rise of the Professional Classes in Nineteenth-Century England* (London: Weidenfeld and Nicolson, 1966), 41. For a theory of the professions as "monopolies of training" and "monopolies of competence" see Larson, 222–25.

50. Clifford Siskin, *The Work of Writing: Literature and Social Change in Britain, 1700–1830* (Baltimore: The Johns Hopkins University Press, 1998), 103–09.

51. Charles Edmonds, ed., *Poetry of the Anti-Jacobin* (London: Sampson, 1890), 147.

52. Elizabeth Hamilton, *Memoirs of Modern Philosophers*, 3 vols. (Bath: G. G. and J. Robinson, 1800), 1:35.

53. Hamilton, *Memoirs*, 1:101.

54. For more on this satirical strategy, see Maurice Crosland, "The Image of Science as a Threat: Burke versus Priestley and the 'Philosophic Revolution,' " *British Journal for the History of Science* 20 (1987): 277–307. It is also possible that British satirists might have been aware of the association of Mesmeric "animal magnetism" with radical politics in France. This tradition is discussed in Robert Darnton, *Mesmerism and the End of the Enlightenment* (Cambridge: Harvard University Press, 1968), 83–105, 164.

55. Elizabeth Hamilton, *Translation of the Letters of a Hindoo Rajah*, 2 vols. (London: G. G. & J. Robinson, 1796), 2:212–13.

56. Hamilton, *Memoirs*, 1:252.

57. Hamilton, *Memoirs*, 1:320–21.

58. Hamilton, *Memoirs*, 1:325.

59. Stephen Greenblatt, *Shakespearean Negotiations: The Circulation of Social Energy in Renaissance England* (Berkeley: University of California Press, 1988), 29–30. This book's title and first chapter incidentally demonstrate that a metaphor of social energy continues to inform late-twentieth-century writing. The metaphor plays a different role in contemporary literary criticism

than it did in the 1790s; see Ted Underwood, "Romantic Historicism and the Afterlife," *PMLA* 117 (2002): 237–51.

60. Frank J. Messman, *Richard Payne Knight: The Twilight of Virtuosity* (Paris: Mouton, 1973), 9–21.

61. Richard Payne Knight, *The Progress of Civil Society* (London: G. Nicol, 1796), 55, v.

62. It is repeated, for instance, by Humphry Davy; see *DW* 2:323.

63. Knight, *Progress of Civil Society*, 84.

64. Knight, *Progress of Civil Society*, 87–88.

65. Davy, notebook 20a, 27.

66. Davy, notebook 22a, 29.

67. Knight, *Progress of Civil Society*, 88.

68. Knight, *Progress of Civil Society*, 89–90.

69. Knight, *Progress of Civil Society*, 90–92.

70. Keen, *Crisis of Literature*, 87, 80.

71. David Williams, *Letters on Political Liberty* (London, 1782), quoted in Dickinson, *Liberty and Property*, 228.

72. For a slightly different exploration of this tension, see Underwood, "Productivism," 122–25. There is some evidence that the growing popularity of *energy* actually reduced writers' reliance on approbative terms like *industry* and *diligence*.

Chapter 3   Apollo, God of Enterprise

1. Marilyn Butler, "Romantic Manichaeism: Shelley's 'On the Devil, and Devils' and Byron's Mythological Dramas," *The Sun is God: Painting, Literature, and Mythology in the Nineteenth Century*, ed. J. B. Bullen (Oxford: Oxford University Press, 1989), 15.

2. Charles-François Dupuis, *Abrégé de l'origine de tous les cultes* (1795; repr., Paris: Bossange, 1820), 96. See also William Drummond, *The Œdipus Judaicus* (London: Valpy, 1811), 142–43, 160–61. For a Christian version of this argument, identifying sun-worship as the original form of paganism, but explaining it as a corruption of the even older "true" religion, see Jacob Bryant, *A New System, or an Analysis of Ancient Mythology*, 3 vols. (1775–1776; repr., New York: Garland, 1979), especially 1:1–2, 1:305–17.

3. Butler, "Romantic Manichaeism," 13–34.

4. Friedrich Hölderlin, "Patmos," *Poems and Fragments*, trans. Michael Hamburger (London: Anvil, 1994), 482–507.

5. Samuel Taylor Coleridge, "Religious Musings," *Complete Poetical Works*, ed. E. H. Coleridge, 2 vols. (Oxford: Oxford University Press, 1968), 1:108–25.

6. John Milton, *Paradise Lost*, ed. Scott Elledge (New York: W. W. Norton, 1975), 56–57 (3:1–5).

7. Thomas Moore, *Poetical Works*, 10 vols. (London: Longmans, 1840–1841), 4:253.

8. Isaac Newton, *Opticks; or, A Treatise of the Reflections, Refractions, Inflections and Coloms of Light* (New York: Dover, 1952), 401.

9. P. M. Heimann and J. E. McGuire, "Newtonian Forces and Lockean Powers: Concepts of Matter in Eighteenth-Century Thought," *Historical Studies in the Physical Sciences* 3 (1971): 233–306.

10. G. W. F. Hegel, *The Philosophy of History*, trans. J. Sibree (New York: Dover, 1956), 103.

11. Mary Wollstonecraft, *A Vindication of the Rights of Woman*, ed. Carol H. Poston (New York: Norton, 1975), 52, 57.

12. Wollstonecraft, *A Vindication*, 57.

13. J. G. A. Pocock, *Virtue, Commerce, and History: Essays on Political Thought and History, Chiefly in the Eighteenth Century* (Cambridge: Cambridge University Press, 1985), 240.

14. James Thomson, "Liberty," *The Complete Poetical Works of James Thomson*, ed. J. Logic Robertson (London: Oxford University Press, 1908), 395 (5:121–23).

15. Thomson, "Liberty," 396 (5:133–36).

16. J. G. A. Pocock, *The Machiavellian Moment: Florentine Political Thought and the Atlantic Republican Tradition* (Princeton: Princeton University Press, 1975), 446–48.

17. Thomson, "Liberty," 396 (5:144–45).

18. Thomson, "Liberty," 408, 357 (5:553–54, 4:11–12).

19. Hermann Boerhaave, *Elements of Chemistry, Being the Annual Lectures of Herman Boerhaave, M. D.* (London: J. & J. Pemberton, 1735), 1:47.

20. Thomson, "Liberty," 373 (4:574–76).

21. Thomson, "Liberty," 390–91 (4:1163–69).

22. W. K. Wimsatt, "The Structure of Romantic Nature Imagery," *The Verbal Icon* (n. p.: University of Kentucky, 1954), 109.

23. Jerome McGann, *The Poetics of Sensibility* (Oxford: Clarendon Press, 1996), 113.

24. The similarity lies in Marx's early conception of alienation as alienation from nature. See Karl Marx, *Economic and Philosophic Manuscripts of 1844* (Moscow: Foreign Languages Publishing House, 1961), 74.

25. Wollstonecraft, *A Vindication*, 140–41.

26. See, for instance, Humphry Davy, "On the Credulity of Mortals," notebook 13f, ms., Humphry Davy Papers, The Royal Institution, London, 24–28.

27. Samuel Taylor Coleridge, Letter 27c/6, ms., Humphry Davy Papers. This letter is also printed in Samuel Taylor Coleridge, *Collected Letters*, ed. E. L. Griggs, 6 vols. (Oxford: Oxford University Press, 1956), 1:630. Griggs has "grandly" instead of "gravely," which I think is a transcription error. The note on p. 630 also has the wrong title for Davy's poem.

28. John Davy, *Memoirs of the Life of Sir Humphry Davy*, 2 vols. (London: Longman, 1836), 1:389–90.

29. Clement Carlyon, *Early Years and Late Reflections*, vol. 1 (London: Whittaker, 1858), 234–35.

30. The notebook version has "their," but "his," from Carlyon's text, makes better sense, since the singular head can hardly belong to the planets and must belong to the Spinosist. The manuscript confusion is nonetheless interesting,

and perhaps symptomatic of an uncertainty in the poem as to whether a specific person, or existence in general, is being described.

31. Davy, notebook 13c, ms., Humphry Davy Papers, 7–10.

32. James Hutton, *Dissertations on Different Subjects in Natural Philosophy* (Edinburgh: A. Strahan and T. Cadell, 1792), iii, ix.

33. See Alice Jenkins, "Humphry Davy: Poetry, Science, and the Love of Light," *1798: The Year of the Lyrical Ballads*, ed. Richard Cronin (Houndmills: Macmillan, 1998), 133–50.

34. See, for instance, Davy's poem titled "1799," notebook 13c, 44.

35. Dewhurst Bilsborrow, "To Erasmus Darwin, on His Work Entitled Zoonomia," *Zoonomia; or, the Laws of Organic Life*, by Erasmus Darwin, 2 vols. (London: J. Johnson, 1794), 1:vii-viii.

36. Humphry Davy, "The Sons of Genius," *The Poetry of Humphry Davy*, ed. Alison Pritchard (Penwith: Penwith District Council, 1978), 2–4.

37. Samuel T. Coleridge, Letter 27c/6, ms., Humphry Davy Papers.

38. See also Adam Walker, *A System of Familiar Philosophy in Twelve Lectures*, 2 vols. (London: Printed for the author, 1802), 1:23, 2:95–98.

39. John Davy, *Memoirs*, 2:368.

40. Henry Boyd, "The Helots," *Poems, Chiefly Dramatic and Lyric* (Dublin: Graisberry and Campbell, 1793), ii.

41. Boyd, "The Helots," 39.

42. Boyd, "The Helots," 40.

43. Boyd, "The Helots," 12, 108.

44. Walter Burkert, *Greek Religion*, trans. John Raffan (Cambridge: Harvard University Press, 1985), 145–48.

45. J. H. (John Harington), *The Odes and Epodon of Horace* (London: W. Crooke, 1684), 25.

46. William Henry Ireland, "Invocation to Genius," *Neglected Genius* (London: Printed by W. Wilson for George Cowrie and Co., 1812), 2.

47. William Henry Ireland, "Fragment," *Rhapsodies* (London: Longman and Rees, 1803), 74.

48. John Keats, *The Poems of John Keats*, ed. Jack Stillinger (Cambridge: Harvard University Press, 1978), 92.

49. Ian Jack, *Keats and the Mirror of Art* (Oxford: Oxford University Press, 1967), 180.

50. John Keats, letter to John Taylor, February 27, 1818, *The Letters of John Keats*, ed. Hyder Edward Rollins, 2 vols. (Cambridge, Mass.: Harvard University Press, 1958), 1:238–39.

51. Keats, *Poems*, 227, 70.

52. Keats, *Poems*, 75.

53. Homer, *Odyssey*, trans. Richmond Lattimore (New York: Harper, 1967), 27.

54. Neil H. Hertz, "Wordsworth and the Tears of Adam," *Studies in Romanticism* 7 (1967): 18.

55. David V. Erdman, *The Illuminated Blake* (New York: Doubleday, 1974), 174.

56. Erdman, *Illuminated Blake*, 181. On the redness of the sun, see Stuart Peterfreund, *William Blake in a Newtonian World: Essays on Literature as Art and Science*, Series for Science and Culture (Norman: University of Oklahoma Press, 1998), 160–61.

57. Northrop Frye, *Fearful Symmetry: A Study of William Blake* (Princeton: Princeton University Press, 1947), 140.

58. See Mark Lussier, *Romantic Dynamics: The Poetics of Physicality* (New York: St. Martin's Press, 2000), 47–63.

59. Morris Eaves, *The Counter-Arts Conspiracy: Art and Industry in the Age of Blake* (Ithaca: Cornell University Press, 1992), 175–253.

60. Eaves, *Counter-Arts Conspiracy*, 169–70.

61. Michael Ferber describes *Jerusalem's* account of alienated labor in *The Social Vision of William Blake* (Princeton, NJ: Princeton University Press, 1985), 131. See also David Punter, "Blake: Creative and Uncreative Labor," *Studies in Romanticism* 16 (1977): 535–61, esp. 549.

62. W. J. T. Mitchell, "Poetic and Pictorial Imagination in Blake's *The Book of Urizen*," *Eighteenth Century Studies* 3 (1969): 103.

63. Marx, *Economic and Philosophic Manuscripts of 1844*, 70.

64. Marx, *Economic and Philosophic Manuscripts of 1844*, 73–75.

Chapter 4   Cowper's Spontaneous *Task*

1. Dustin Griffin has suggested that *The Task* "redefines georgic" by "redefining labor . . . as a virtually spiritual activity, and . . . shifting attention from the public sphere to the private." Richard Feingold, on the other hand, feels that the poem's failure is the real riddle to be explained; he discusses *The Task* as a sensitive, but essentially doomed, effort to fit late-eighteenth-century society into an antiquated poetic mode. Dustin Griffin, "Redefining Georgic: Cowper's Task," *English Literary History* 57.4 (1990): 876. Richard Feingold, *Nature and Society* (New Brunswick: Rutgers University Press, 1978): 121–92.

2. John Dyer, *The Fleece, Minor Poets of the Eighteenth Century*, ed. H. Fausset (New York: E. P. Dutton, 1930), 3:22–26.

3. James King, *William Cowper: A Biography* (Durham: Duke University Press, 1986), 155.

4. Francis Jeffrey, rev. of *The Dramatic Works of John Ford*, *Contributions to the Edinburgh Review*, ed. Henry Weber, 4 vols. (London: Longmans, 1944), 2:294.

5. The survey covered middle-class memoirs, diaries, letters, and family records in the East and West Midlands between 1780 and 1850. "Cowper is easily the most quoted writer in the records from both areas . . . usually occupying a place of honour." The authors indicate that this remains true throughout the period of their study. Leonore Davidoff and Catherine Hall, *Family Fortunes: Men and Women of the English Middle Class 1780–1850* (Chicago: University of Chicago Press, 1987), 157, 157n.

6. Martin Priestman has similarly argued that Cowper's *Task* struggles to reconcile a Christian work ethic with a tradition of retirement poetry. Martin Priestman, *Cowper's Task: Structure and Influence* (Cambridge: Cambridge University Press, 1983), 96.

7. Feingold, *Nature and Society*, 5–7. See also C. A. Moore, "Whig Panegyric Verse, 1700–1760," *PMLA* 41 (1926): 362–401.

8. Virgil, *The Georgics*, trans. L. P.Wilkinson (NewYork: Penguin, 1982), 1.197–201.

9. Virgil, *The Georgics*, 1.145–6.

10. Dwight L. Durling, *Georgic Tradition in English Poetry* (New York: Columbia University Press, 1935), 33–35.

11. John Milton, *Paradise Lost*, ed. Scott Elledge (NewYork:W.W. Norton, 1975), 94 [4:618–22].

12. Ben Jonson, "To Penshurst," *Poems*, ed. I. Donaldson (London: Oxford University Press, 1975), 88.

13. Feingold, *Nature and Society*, 33.

14. RaymondWilliams, *The Country and the City* (NewYork: Oxford University Press, 1973), 29–31.

15. In saying this, I echo John Chalker's observation that Thomson's "sensitively serious eighteenth-century imitation of the *Georgics*" was both more impor- tant for and more representative of its time than more formal imitations such as John Philips's *Cyder* (1708) or John Dyer's *The Fleece* (1757) that retain the Virgilian didactic structure. John Chalker, *The English Georgic: A Study in the Development of a Form* (Baltimore: Johns Hopkins Press, 1969), 92.

16. Durling, *Georgic Tradition*, 43–44. For a broader interpretation of theVirgilian influence on Thomson see Chalker, *The English Georgic*, 90–140.

17. Durling, *Georgic Tradition*, 46.

18. James Thomson, "Summer," *The Complete Poetical Works of James Thomson*, ed. J. L. Robertson (London: Oxford University Press, 1908), 1438–45.

19. Thomson, "Spring," *Complete Poetical Works*, 852–58.

20. John Dyer, *The Fleece*, 3:22–26.

21. "Labour n. s. 1. The act of doing what requires a painful exertion of strength, or wearisome perseverance." Samuel Johnson, *A Dictionary of the English Language* (London: Strahan, 1755). Likewise, Burke's description of the sublime as a kind of inward labor postulates a resistance overcome, since "labour is a surmounting of *difficulties*" (Burke's italics). Edmund Burke, *A Philosophical Inquiry into the Origin of our Ideas of the Sublime and Beautiful*, ed. J. T. Boulton (NewYork: Columbia University Press, 1958), 135.

22. "In agriculture too nature labours along with man; and though her labour costs no expence, its produce has its value, as well as that of the most expen- sive workmen." Adam Smith, *An Inquiry into the Nature and Causes of the Wealth of Nations*, ed. R. H. Campbell, A. S. Skinner, and W. B. Todd, 2 vols. (Oxford: Oxford University Press, 1976), 1:363 (II.v.12).

23. Dyer, *The Fleece*, 3:79–83.

24. Dyer, *The Fleece*, 3:89–95.

25. Feingold, *Nature and Society*, 88.

26. William Cowper, letter to William Unwin, October 10, 1784, *The Letters and Prose Writings of William Cowper*, ed. J. King and C. Ryskamp (Oxford: Oxford University Press, 1981), 285.

27. Priestman, *Cowper's Task*, 10.

28. Priestman, *Cowper's Task*, 3, 14. See also Dorothy Hadley Craven, "Cowper's Use of 'Slight Connection' in The Task: A Study in Structure and Style," (PhD diss., University of Colorado, 1953).

29. My account of Cowper's life is primarily based on King, *William Cowper*, 1–157.
30. King, *William Cowper*, 58.
31. King, *William Cowper*, 59–60.
32. Priestman, *Cowper's Task*, 14–17.
33. William Cowper, "Table Talk," *The Poems of William Cowper*, ed. J. D. Baird and C. Ryskamp, 3 vols. (Oxford: Oxford University Press, 1995), 1:257. The lines are, significantly enough, placed in the mouth of the Whig speaker in a Whig–Tory dialogue.
34. William Cowper, letter to William Unwin, October 20, 1784, *Letters*, 286.
35. William Cowper, letter to William Unwin, October 10, 1784, *Letters*, 285.
36. The assumption that association is the mind's natural mode of movement is drawn from Locke and Hartley. See Priestman, *Cowper's Task*, 12, 22.
37. William Cowper, letter to William Unwin, October 10, 1784, *Letters*, 285.
38. Priestman, *Cowper's Task*, 96.
39. Cowper's *Task* is followed by a long tradition of nineteenth-century works in which the activity of walking has to support this sort of medico-moral argument. Examining Cowper, William Wordsworth, H. D. Thoreau, George Eliot, and John Burroughs, among others, Anne D. Wallace's *Walking, Literature, and English Culture* concludes that the georgic mode is replaced by a new mode she names "peripatetic," which puts the walker in "the ideological space vacated by the farmer," and "represents excursive walking as a cultivating labour capable of renovating both the individual and his society by recollecting and expressing past value." Anne D. Wallace, *Walking, Literature, and English Culture: The Origins and Uses of Peripatetic in the Nineteenth Century* (Oxford: Oxford University Press, 1993), 11.
40. "Tone, n. s. . . . . 5. Elasticity; Power of Extension and Contraction." Samuel Johnson, *Dictionary*.
41. William Gilpin, *Three Essays: On Picturesque Beauty, on Picturesque Travel, and on Sketching Landcape*, 2nd ed. (London: Blamire, 1794), 22–23.
42. Gilpin, *Three Eassays*, 11–12.
43. Wallace, *Walking*, 95–102. Thomas Pfau, *Wordsworth's Profession: Form, Class, and the Logic of Early Romantic Cultural Production* (Stanford, California: Stanford University Press, 1997), 61–82.
44. Samuel Taylor Coleridge, *The Notebooks of Samuel Taylor Coleridge*, ed. Kathleen Coburn, 4 vols. (New York: Pantheon, 1957), 1:1589, 1:785.
45. Samuel Taylor Coleridge, *Complete Poetical Works*, ed. E. H. Coleridge, 2 vols. (Oxford: Oxford University Press, 1968), 1:179, 1:181.

Chapter 5   Wordsworth and the Homelessness of Engines

1. William Wordsworth, letter to Dorothy Wordsworth, April 24, 1812, *The Letters of William and Dorothy Wordsworth: A Supplement of New Letters*, ed. Alan G. Hill (New York: Oxford University Press, 1993), 44.

2. See for instance Raymond Williams, *The Country and the City* (New York: Oxford University Press, 1973), 127–133. Also Kurt Heinzelman, *The Economics of the Imagination* (Amherst: University of Massachusetts Press, 1980), 196–233.

3. William Spence, *Agriculture the Source of the Wealth of Britain* (1808), repr. in William Spence, *Tracts on Political Economy* (London: Longman, Hurst, Rees, Orme, and Brown, 1822), 100, 122.

4. T. R. Malthus, "Spence on Commerce," *Edinburgh Review* 11 (1808): 429–48.

5. T. R. Malthus, *An Essay on the Principle of Population*, 3rd ed., 2 vols. (London: J. Johnson, 1806), 2:212–13.

6. T. R. Malthus, *The Grounds of an Opinion on the Policy of Restricting the Importation of Foreign Corn* (London: Printed for John Murray and J. Johnson, 1815), 34–36.

7. Alison Hickey points out that Wordsworth's praise of industry in *The Excursion* has affinities with his theory of language. I am indebted to her insightful discussion of the topic, though I do not share her sense that the Wanderer's admiration is ironic or insincere. See Alison Hickey, *Impure Conceits: Rhetoric and Ideology in Wordsworth's Excursion* (Stanford: Stanford University Press, 1997), 99–101.

8. Wordsworth had many acquaintances and an adequate allowance at Cambridge. Kenneth Johnston, *The Hidden Wordsworth: Poet, Lover, Rebel, Spy* (New York: W. W. Norton, 1998), 111–119. But social unease is a relative condition; it has more to do with the contrast between one's own station and others' than with objective sufficiency of resources. In that relative sense the transition from Hawkshead to Cambridge could have been as disorienting as *The Prelude* implies: Wordsworth was for the first time surrounded by men who were for the most part his social superiors.

9. Stephen Gill, *William Wordsworth: A Life* (Oxford: Oxford University Press, 1989), 35.

10. David Simpson, *Wordsworth's Historical Imagination: The Poetry of Displacement* (New York: Methuen, 1987), 56–78.

11. Charles Rzepka, " 'A Gift that Complicates Employ': Poetry and Poverty in 'Resolution and Independence,' " *Studies in Romanticism* 28 (1989): 225–47. See also Toby R. Benis, *Romanticism on the Road: The Marginal Gains of Wordsworth's Homeless* (New York: St. Martin's, 2000).

12. Thomas Malthus, *An Inquiry into the Nature and Progress of Rent, and the Principles by Which It is Regulated* (London: Printed for John Murray and J. Johnson, 1815), 20–21.

13. David Ricardo, *On the Principles of Political Economy and Taxation, Works and Correspondence*, ed. Pierro Sraffa and M. H. Dobb, 10 vols. (Cambridge: Cambridge University Press, 1966), 1:75.

14. John Milton, *Paradise Lost*, ed. Scott Elledge (New York: W. W. Norton, 1975), 281 (12:646–47).

15. Alan Liu, *Wordsworth: The Sense of History* (Stanford: Stanford University Press, 1989), 342.

16. As Philip Connell has pointed out, early-nineteenth-century readers of *The Excursion* did not seem to hear the irony that critics have recently located in this passage. Connell also persuasively stresses the connection between Wordsworth's growing enthusiasm for industry after 1809 and his "growing commitment to a rhetoric of patriotism and national solidarity." Philip Connell, *Romanticism, Economics, and the Question of 'Culture,'* (Oxford: Oxford University Press, 2001), 168-69.

17. William Wordsworth, *The Ruined Cottage and The Pedlar*, ed. James Butler (Ithaca: Cornell University Press, 1979), 73, 273.

18. Adam Smith, *An Inquiry into the Nature and Causes of the Wealth of Nations*, ed. R. H. Campbell, A. S. Skinner, and W. B. Todd, 2 vols. (Oxford: Oxford University Press, 1976), 1:363–64 (II.v.12).

19. Smith, *An Inquiry*, 356 (II.iv.13).

20. See e.g. Malthus, *Grounds of an Opinion*, 34–36.

21. Alexander Hamilton, *The Reports of Alexander Hamilton*, ed. Jacob E. Cooke (New York: Harper, 1964), 121.

22. Jean-Baptiste Say, *Traité d'Économie Politique* (Paris: Deterville, 1803), 39–40.

23. Malthus, *Grounds of an Opinion*, 34–36.

24. Ricardo, *Works*, 1:76n.

25. "I shall never forget Mr. Bolton's expression to me:'I sell here, Sir, what all the world desires to have—POWER.' " James Boswell, *Boswell's Life of Johnson*, ed. G. B. Hill and L. F. Powell, vol. 2 (Oxford: Oxford University Press, 1971), 459.

26. Annabel Patterson, "Wordsworth's Georgic: Genre and Structure in *The Excursion*," *The Wordsworth Circle* 9 (1978): 148, 147, 153.

27. Simpson, *Wordsworth's Historical Imagination*, 51–54, 200–212.

28. Humphry Davy, notebook 13c, ms., Humphry Davy Papers, The Royal Institution, London, 7–10.

29. Smith, *An Inquiry*, 1:363–64.

30. Ricardo, *Works*, 1:70, 1:69.

31. Ricardo, *Works*, 1:75.

32. Thomas Carlyle, *The Works of Thomas Carlyle*, 30 vols. (New York: Scribner's, 1896), 1:56.

33. Simpson, *Wordsworth's Historical Imagination*, 48–49.

34. John Muir, *My First Summer in the Sierra* (Boston: Houghton Mifflin, 1911), 176. Compare his similar responses to Douglas squirrels and bears: 96, 138.

35. Henry David Thoreau, *Walden and Resistance to Civil Government*, ed. William Rossi and Owen Thomas (New York: W. W. Norton, 1992), 2–3, 203–06. For an account of Thoreau's identification with the productive energy revealed in thawing sand and clay, see Eric G. Wilson, *The Spiritual History of Ice: Romanticism, Science, and the Imagination* (New York: Palgrave, 2003), 54–58.

36. Leo Marx, *The Machine in the Garden: Technology and the Pastoral Ideal in America* (Oxford: Oxford University Press, 2000), 229–39.

37. Marx, *The Machine in the Garden*, 162, 231.

Chapter 6 Sunlight and the Reification of Culture

1. Ian Jack, *Keats and the Mirror of Art* (Oxford: Oxford University Press, 1967), 180.
2. John Keats, "Sleep and Poetry," *The Poems of John Keats*, ed. Jack Stillinger (Cambridge, MA: Harvard University Press, 1978), 74.
3. William Hazlitt, *The Complete Works*, ed. P. P. Howe, 21 vols. (London: J. M. Dent, 1930), 5:3, 5:1, 5:3.
4. Friedrich Schlegel, *Dialogue on Poetry and Literary Aphorisms*, trans. Ernst Behler and Roman Struc (University Park: Pennsylvania State University Press, 1968), 54.
5. Marx's "vergegenständliche Arbeit" is sometimes translated as "reified labor," but just as often in English it becomes "objectified" or "embodied" labor. For instance, "Ein Gebrauchswert oder Gut hat also nur einen Wert, weil abstrakt menschliche Arbeit in ihm vergegenständlicht oder materialisiert ist" is translated as "A use value, or useful article, therefore, has value only because abstract human labor is objectified or materialised in it." Karl Marx, *Capital: A Critique of Political Economy*, trans. Ben Fowkes, vol. 1 (New York: Vintage, 1977), 129.
6. Georg Lukács, "Reification and the Consciousness of the Proletariat," *History and Class Consciousness*, trans. Rodney Livingstone (1922; Cambridge, MA: MIT Press, 1971), 83–222. Returning to this topic forty-five years later, Lukács attempted to distinguish the mere "objectification" that creates nouns from the "reification" that characterizes capitalist ideology (*History and Class Consciousness*, xxiv). I think his later definition of "reification" still conflates a critique of capitalism with a general critique of figurative language. Reification (or objectification) can give form to many different ideological impulses. Commodity fetishism is a capitalist example, but I would not conclude from that example that capitalist society has a special affinity for reification. For further discussion of this topic, see chapter 7.
7. Hannah Arendt, quoted in Jerome Christensen, *Lord Byron's Strength: Romantic Writing and Commercial Society* (Baltimore: Johns Hopkins University Press, 1993), xvi.
8. Christensen, *Lord Byron's Strength*, 5, xxii.
9. Humphry Davy, "The Life of the Spinosist," notebook 13c, ms., Humphry Davy Papers, The Royal Institution, London, 7–10.
10. Thomas Holcroft, *The Adventures of Hugh Trevor*, ed. Seamus Deane (London: Oxford University Press, 1973), 3, 97.
11. William Godwin, *Caleb Williams*, ed. David McCracken (London: Oxford University Press, 1970), 218–19.
12. Keats, "Isabella; or, the Pot of Basil," *Poems*, 249.
13. John Clare, "The Dawning of Genius," *The Early Poems of John Clare: 1804–1822*, ed. Eric Robinson and David Powell, vol. 1 (Oxford: Oxford University Press, 1989), 451. For supporting evidence see Anne D. Wallace, *Walking, Literature, and English Culture: The Origins and Uses of Peripatetic in the*

Nineteenth Century (Oxford: Oxford University Press, 1993), 105–07. For a fuller account of the diversity of eighteenth-century plebeian poetry see William J. Christmas, *The Lab'ring Muses: Work, Writing, and the Social Order in English Plebeian Poetry, 1730–1830* (Newark: University of Delaware Press, 2001), 157–234.

14. Mary Wollstonecraft, *A Vindication of the Rights of Woman*, ed. Carol H. Poston (New York: Norton, 1975), 57.

15. Letitia Elizabeth Landon [L. E. L.], "Lines of Life," *Selected Writings*, ed. Jerome McGann and Daniel Riess (Peterborough, Ontario: Broadview, 1997), 114.

16. Pierre Bourdieu, "What Makes a Social Class? On the Theoretical and Practical Existence of Groups," *Berkeley Journal of Sociology* 32 (1987): 1–17.

17. Randal Johnson, "Editor's Introduction," *The Field of Cultural Production: Essays on Art and Literature*, by Pierre Bourdieu (New York: Columbia University Press, 1993), 7.

18. Pierre Bourdieu, *The Field of Cultural Production: Essays on Art and Literature* (New York: Columbia University Press, 1993), 37–43.

19. Raymond Williams, *Culture and Society: 1780–1950* (New York: Columbia University Press, 1983), 30–48, 110–29. See also Raymond Williams, *Keywords: A Vocabulary of Culture and Society* (New York: Oxford University Press, 1983), 87–93.

20. Trevor Ross, " 'Pure Poetry': Cultural Capital and the Rejection of Classicism," *Modern Language Quarterly* 58 (1997): 444–45. See also John Guillory, *Cultural Capital* (Chicago: University of Chicago Press, 1993), 124–33.

21. Williams, *Keywords*, 88.

22. George Gordon, Lord Byron, *Letters and Journals*, ed. Leslie A. Marchand (London: Murray, 1979), 46.

23. Marilyn Butler, *Romantics, Rebels, and Reactionaries: English Literature and Its Background 1760–1830* (Oxford: Oxford University Press, 1990), 113–37.

24. See Alan Bewell, "Jefferson's Thermometer: Colonial Biogeographical Constructions of the Climate of America," *Romantic Science: The Literary Forms of Natural History*, ed. Noah Heringman (Albany, NY: SUNY Press, 2003), 111–38.

25. Davy, notebook 13c, 7–10.

26. Jerome McGann, "Byron, Mobility, and the Poetics of Historical Ventriloquism," *Byron and Romanticism*, by Jerome McGann, ed. James Soderholm (Cambridge: Cambridge University Press, 2002), 36–52.

27. John Brown, *The Elements of Medicine*, ed. Thomas Beddoes, 2 vols. (London: J. Johnson, 1795), 1:xvii. See chapter 1, "Light as fuel for body and mind," for a fuller discussion of Brown.

28. Benjamin Waterhouse, *On the Principle of Vitality* (Boston: Fleet, 1790). Joseph Trent, *An Inquiry Into the Effects of Light in Respiration* (Philadelphia: Way and Graff, 1800). T. Gale, *Electricity, or Ethereal Fire* (Troy: Moffitt & Lyon, 1802), 15, 89.

29. Thomas Moore, "Song of a Hyperborean," *Poetical Works*, 10 vols. (London: Longmans, 1840–1841), 5:253.

30. In taking this theme as central to Canto IV, I follow Jerome McGann, *Fiery Dust: Byron's Poetic Development* (Chicago: University of Chicago Press, 1968), 130–32.

31. Walter Burkert, *Greek Religion*, trans. John Raffan (Cambridge: Harvard University Press, 1985), 145–48.

32. In this period, the catchphrase "light and life" (or vice versa) becomes very common, indeed almost inescapable, in poems about the sun. It is one of the most noticeable tics of Shelley's style, but see also, for instance, Anna Seward, "Ode to the Sun," *The Poetical Works of Anna Seward*, ed. Walter Scott, 3 vols. (Edinburgh: J. Ballantyne, 1810), 2:49–52. Charlotte Smith, "To the Sun," *Elegiac Sonnets*, 8th ed., 2 vols. (London: T. Cadell and W. Davies, 1797–1800), 2:30. Robert Southey, "The Peruvian's Dirge over the Body of his Father" (1799), *Poetical Works*, 10 vols., (London: Longmans, 1838), 2:207–9. John Clare, "The Sun," *The Later Poems of John Clare*, ed. Eric Robinson and David Powell, 2 vols. (Oxford: Oxford University Press, 1984), 1:353–54.

33. Harold Bloom, *The Visionary Company: A Reading of English Romantic Poetry* (New York: Doubleday, 1963), 258–59.

34. Friedrich Schiller, *On the Aesthetic Education of Man: In a Series of Letters*, ed. and trans. Elizabeth M. Wilkinson and L. A. Willoughby (Oxford: Oxford University Press, 1967), 43.

35. See in particular Williams, *Culture and Society*, 30–48. For a different (but I believe quite compatible) approach to the same problem, see Geoffrey H. Hartman, "Romanticism and Anti-Self-Consciousness," *Romanticism and Consciousness*, ed. Harold Bloom (New York: W. W. Norton, 1970), 46–56.

36. Jeremiah Holmes Wiffen, "Aspley Wood," *Aonian Hours, and Other Poems* (London: Longman, Hurst, Rees, Orme, and Brown, 1819), vi, 78.

37. Wiffen, "Aspley Wood," 85, 79–85.

38. Wiffen, "Aspley Wood," 14, 15.

39. Byron, *Letters and Journals*, 46.

40. Bryan Procter Waller ("Barry Cornwall"), "The Girl of Provence," *The Flood of Thessaly, The Girl of Provence, and Other Poems* (London: Henry Colburn, 1823), 88, 114.

41. Waller, "The Girl of Provence," 93, 115, 117, 119.

42. Waller, "The Girl of Provence," 89.

43. Mary Shelley, "Midas," *Proserpine and Midas*, ed. A Koszul (London: Humphrey Milford, 1922), 55.

44. Earl Wasserman, *Shelley: A Critical Reading* (Baltimore: Johns Hopkins University Press, 1971), 49, 55.

45. Particle size has this effect because light goes through "fits of easy reflection and easy transmission." Isaac Newton, *Opticks; or, A Treatise of the Reflections, Refractions, Inflections and Colours of Light* (New York: Dover, 1952), 281–85.

46. Adam Walker, *A System of Familiar Philosophy in Twelve Lectures*, 2 vols. (London: Printed for the author, 1802), 2:102–03.

47. Andrew Amos, "Shelley and His Contemporaries at Eton," *Athenaeum*, (April 15, 1848), 390. The connection between Walker and Shelley is also

discussed in Peter Butter, *Shelley's Idols of the Cave* (Edinburgh: Edinburgh University Press, 1954), 136, 140–48.

48. Humphry Davy, *Elements of Chemical Philosophy* (London: J. Johnson, 1812), 213. Mary Shelley, *Journal*, ed. F. L. Jones (Norman: University of Oklahoma Press, 1947), 67.

49. *British Encyclopedia; or, Dictionary of the Arts and Sciences* (London: Longman, Hurst, Rees, 1809–1811), s.v. "astronomy."

50. James Hutton, *A Dissertation on the Philosophy of Light, Heat, and Fire* (Edinburgh: Cadell, 1794), 323. See also Davy, *Elements* of chemical philosophy (London: J. Johnson), 213 and *British Encyclopedia*, s.v. "light."

51. Williams, *Culture and Society*, 42–43.

52. A description of the sun as the "world's eye" can be found both in Ovid and in Milton, but not with this implication. See Ovid, *Metamorphoses*, trans. Rolfe Humphries (Bloomington: Indiana University Press, 1955), 4:226–28; and John Milton, *Paradise Lost*, ed. Scott Elledge (New York: Norton, 1975), 108 (5:171–72).

53. Jean Senebier, *Mémoires physico-chymiques sur l'influence de la lumière solaire pour modifier les êtres des trois règnes de la nature, et sur-tout ceux du règne végétal*, 3 vols. (Geneva: Chiral, 1782), 1:viii. See chapter 1, "The identity of agency and appearance," for a fuller discussion of Senebier.

54. Percy Bysshe Shelley, *The Letters of Percy Bysshe Shelley*, ed. Frederick L. Jones, 2 vols. (Oxford: Oxford University Press, 1964), 2:271.

55. Shelley, *Letters*, 2:273. Shelley's cancellations are given in square brackets and italics.

56. M. H. Abrams, *The Mirror and the Lamp: Romantic Theory and the Critical Tradition* (New York: Norton, 1958), 130.

57. Milton, *Paradise Lost*, 73 (3:608–09).

58. Angela Leighton, *Shelley and the Sublime: An Interpretation of the Major Poems* (Cambridge: Cambridge University Press, 1984), 45–47.

Chapter 7   Energy Becomes Labor: The Role of Engineering Theory

1. John F. W. Herschel, *A Treatise of Astronomy* (London: Longman, 1833), 211.

2. Thomas Carlyle, *The Works of Thomas Carlyle*, 30 vols. (New York: Scribner's, 1896), 1:56.

3. Hannah Arendt, *The Human Condition* (Chicago: University of Chicago Press, 1958), 79–93.

4. *GW* 5:154. See "Energy versus 'mechanical and daily labor' " in chapter 2.

5. Though the social situation was different in Germany (with relatively less idealization of the middle classes, and more idealization of the peasant's relationship to the soil), G. W. F. Hegel reproduces the same distinction in his Jena lectures of 1805–1806. "Concrete labor"—performed by peasants—is "the elementary labor, the substantial sustenance, the crude basis of the whole. . . . [It] leaves a fertile mud and then is spent, having produced no

[meaningful] work." But "the labor of the Bürger class is the abstract labor of individual handicrafts. . . . The self has [thus] gained independence from the earth." G. W. F. Hegel, *Hegel and the Human Spirit: A Translation of the Jena Lectures on the Philosophy of Spirit (1805–1806)*, trans. Leo Rauch (Detroit: Wayne State University Press, 1983), 164–65.

6. *GW* 5:154; see also *GW* 3:439–40.

7. Thomas Love Peacock, "Ahrimanes," *Works*, 10 vols. (London: Constable, 1924–1934), 7:269, 7:274.

8. Dror Wahrman, *Imagining the Middle Class: The Political Representation of Class in Britain* (Cambridge,: Cambridge University Press, 1995), 214–17.

9. Ernest Solvay, *Principes d'Orientation Sociale* (Bruxelles: Misch et Thron, 1904), 33.

10. Georg Lukács, "Reification and the Consciousness of the Proletariat," *History and Class Consciousness*, trans. Rodney Livingstone (Cambridge, MA: MIT Press, 1971), 83–222.

11. Anson Rabinbach, *The Human Motor: Energy, Fatigue, and the Origins of Modernity* (New York: Basic Books, 1990), 17, 206–88.

12. Rabinbach, *The Human Motor*, 69–83.

13. For a study of quantification as a "technology of distance" that permits communication in contexts where interpersonal trust is no longer adequate, see Theodore M. Porter, *Trust in Numbers: The Pursuit of Objectivity in Science and Public Life* (Princeton: Princeton University Press, 1995), ix.

14. Jean-Baptiste Say, *Traité d'Economie Politique* (Paris: Deterville, 1803), 39–40.

15. Attempts to explain the superior productivity of the high-pressure steam engine were eventually important to the development of thermodynamics, as D. S. L. Cardwell has shown, but they were not particularly relevant to this question of mechanics, for which the overshot waterwheel always provided the primary model. For a discussion of the importance of high-pressure steam see D. S. L. Cardwell, *From Watt to Clausius: The Rise of Thermodynamics in the Early Industrial Age* (1971; Ames: Iowa State University Press, 1989), 150–238.

16. "Mechanics," *Encyclopædia Britannica* (Edinburgh: 1810), 51.

17. Understood in twentieth-century terms, this assumption takes the form of an insistence on defining motive power not as a quantity expressible in terms of mass and measurable velocity, but as some such quantity multiplied by $1/t$. Momentum is measured as mv, but an agent's "force" is measured as mv/t. Work and energy are both proportional to mv^2, but an agent's "power" is proportional to mv^2/t.

18. John Smeaton, *An Experimental Enquiry Concerning the Natural Powers of Water and Wind to Turn Mills, and Other Machines, Depending on a Circular Motion* (London: Royal Society, 1760), 8.

19. John Smeaton, *An Experimental Inquiry . . . and an Experimental Examination of the Quantity and Proportion of Mechanic Power Necessary to be Employed in Giving Different Degrees of Velocity to Heavy Bodies from a State of Rest* (1776; rpt. London: Taylor, 1796), 90.

20. Terry S. Reynolds, *Stronger Than a Hundred Men: A History of the Vertical Water Wheel* (Baltimore: Johns Hopkins University Press, 1983), 226.

21. Peter Lundgreen, "Engineering Education in Europe and the U.S.A, 1750–1930:The Rise to Dominance of School Culture and the Engineering Professions," *Annals of Science* 47 (1990): 36–47. (According to Lundgreen, the delayed development of "school culture" in Anglo-American engineering is explained by the fact that, in the Anglo-American context, professional certifications in general were not received through the state.)

22. Lundgreen, *Engineering Education*, 47. There are several clear instances of experimental investigations that produced practical results—Watt's work on the steam-engine and Smeaton's work on the water wheel are two classic examples. But there is little evidence of an effort in Britain to formalize engineering problems in general mathematical terms. Alan Q. Morton has tried to make a case for a closer connection between natural philosophy and practical mechanics in eighteenth-century Britain, in regard to Desaguliers's "maximum machine," but in fact his article turns up only parallels that he admits may be fortuitous. Alan Q. Morton, "Concepts of Power: Natural Philosophy and the Uses of Machines in Mid-Eighteenth-Century London," *British Journal for the History of Science* 28 (1995): 75.

23. Charles Gillispie, *Lazare Carnot, Savant* (Princeton: Princeton University Press, 971), 82.

24. Gillispie, *Lazare Carnot*, 32.

25. Lazare N. M. Carnot, *Principes fondamentaux de l'equilibre et du mouvement* (Paris: Bachelier, 1803), 244–45.

26. Carnot, *Principes*, 246–47.

27. J. N. P. Hachette, *Traité élémentaire des machines* (Paris: Klostermann, 1811), 1.

28. Hachette, *Traité élémentaire*, 126.

29. G. G. Coriolis, *Du calcul de l'effet des machines, ou considérations sur l'emploi des moteurs et sur leur évaluation, pour servir d'introduction à l'étude spéciale des machines* (Paris: Carilian-Goery, 1829), 17.

30. Coriolis, *Du calcul de l'effect des machines*, 27–28.

31. Coriolis, *Du calcul de l'effet des machines*, 31.

32. Peter Ewart, "On the Measure of Moving Force," *Memoirs of the Literary and Philosophical Society of Manchester* 2nd series, 2 (1813): 105–258.

33. Charles Babbage, "On the General Principles which Regulate the Application of Machinery to Manufacture and the Mechanical Arts," *Encyclopaedia Metropolitana*, vol. 8 (1829; London: B. Fellowes et al., 1845), 6.

34. Babbage, "On the General Principles," 7.

35. The early-nineteenth-century importance of this model of nature as a dynamic equilibrium oscillating around a stable balance point has been discussed ably by M. Norton Wise (with the collaboration of Crosbie Smith). Wise stresses that the model is firmly grounded in both the natural philosophy and political economy of the early nineteenth century. M. Norton Wise, with the collaboration of Crosbie Smith, "Work and Waste: Political Economy and Natural Philosophy in Nineteenth-Century Britain (I)," *History of Science* 27 (1989): 263–301.

36. Charles Babbage, *On the Economy of Machinery and Manufactures* (London: Charles Knight, 1832), 317.

37. Nassau W. Senior, *An Outline of the Science of Political Economy* (London: Clowes, 1836), 69.
38. Joseph A. Schumpeter, *History of Economic Analysis*, ed. Elizabeth Boody Schumpeter (New York: Oxford University Press, 1954), 560.
39. Senior, *An Outline*, 58.
40. Carlyle, *Works*, 1:157.
41. Carlyle, *Works*, 10:196.
42. Carlyle, *Works*, 2:59.
43. Carlyle, *Works*, 2:63.
44. Carlyle, *Works*, 2:68.
45. Carlyle, *Works*, 2:59–60.
46. Leo Marx, *The Machine in the Garden: Technology and the Pastoral Ideal in America* (Oxford: Oxford University Press, 2000), 169–90.
47. Carlyle, *Works*, 1:56.
48. Carlyle, *Works*, 10:196, 10:199.
49. Karl Marx, *Capital: A Critique of Political Economy*, trans. Ben Fowkes (New York: Vintage, 1977), 283.

Chapter 8   Productivism and the Popularization of the First Law of Thermodynamics

1. William Thomson, "On a Universal Tendency in Nature to the Dissipation of Mechanical Energy," *Mathematical and Physical Papers*, vol. 1 (Cambridge: Cambridge University Press, 1882), 514. For a contemporary observer's comment on the delayed reception of the second law, see Henry Adams, *The Degradation of the Democratic Dogma* (New York: Macmillan, 1919), 142–43.
2. Hermann von Helmholtz, "On the Conservation of Force; a Physical Memoir," *Scientific Memoirs, Selected from the Transactions of Foreign Academies of Science, and from Foreign Journals: Natural Philosophy*, ed. by John Tyndall and William Francis (London: Taylor & Francis, 1853), 122. William John Macquorn Rankine, "On the General Law of the Transformation of Energy," *Philosophical Magazine* series. 4, 5 (1853): 106.
3. Herbert Spencer, *First Principles* (London: Williams and Norgate, 1862), 258. Michael Faraday, "On the Conservation of Force," *Experimental Researches in Chemistry and Physics* (London: Taylor & Francis, 1859), 447.
4. Edward L. Youmans, "Introduction," *The Correlation and Conservation of Forces* (New York: Appleton, 1865), xli.
5. See Geoffrey Cantor, "Faraday's Search for the Gravelectric Effect," *Physics Education* 26 (1991): 289–93.
6. Spencer, *First Principles*, 253.
7. Anne Gilchrist, "The Indestructibility of Force," *Macmillan's Magazine* 6 (1862): 337.
8. James Hinton, "Force," *Cornhill Magazine* 4 (1861): 414.
9. David Gooding, "Metaphysics versus Measurement: the Conversion and Conservation of Force in Faraday's Physics," *Annals of Science* 37 (1980): 1–29.

10. M. Norton Wise, with the collaboration of Crosbie Smith, "Work and Waste: Political Economy and Natural Philosophy in Nineteenth-Century Britain," *History of Science* 27 (1989): 263–301, 391–449 and 28 (1990): 221–61. See especially 27 (1989): 400–410.

11. Crosbie Smith, *The Science of Energy: A Cultural History of Energy Physics in Victorian Britain* (Chicago: University of Chicago Press, 1998), 100–125.

12. Anson Rabinbach, *The Human Motor: Energy, Fatigue, and the Origins of Modernity* (New York: Basic Books, 1990), 3.

13. Philip Mirowski, *More Heat Than Light: Economics as Social Physics, Physics as Nature's Economics* (Cambridge: Cambridge University Press, 1989), 139–353.

14. Bruce Clarke, *Energy Forms: Allegory and Science in the Era of Classical Thermodynamics* (Ann Arbor: University of Michigan Press, 2001), 59–81.

15. William Thomson, "On the Sources Available to Man for the Production of Mechanical Effect," *Mathematical and Physical Papers*, vol. 1, 510.

16. Herman von Helmholtz, "On the Interaction of Physical Forces," trans. John Tyndall, *The Correlation and Conservation of Forces*, ed. Edward L. Youmans (New York: Appleton, 1865), 240. Justus Liebig, "The Connection and Equivalence of Forces," *The Correlation and Conservation of Forces*, ed. Edward L. Youmans (New York: Appleton, 1865), 397.

17. "Sun Force and Earth Force," *The Popular Science Review* 5 (1866): 327.

18. "The Source of Labour," *Chambers's Journal* series. 4, 3 (1866): 555–56.

19. James B. Russell, "The Labour of the Sunbeams," *Recreative Science* 3 (1862): 60.

20. J. Carpenter, "What We Owe to the Sun," *Once a Week* 15 (1866): 413. See also "Force and Matter," *All the Year Round*, July 21 1866, 35–38; William Thomson and P. G. Tait, "Energy," *Good Words* 3 (1862): 601–07.

21. Spencer, *First Principles*, 282–83.

22. Youmans, "Introduction," xxv, xxvii, xli.

23. J. F. W. Herschel, *A Treatise of Astronomy* (London: Longman, 1833), 211.

24. Adam Walker, *A System of Familiar Philosophy in Twelve Lectures*, 2 vols. (London: Printed for the author, 1802), 2:95, 2:98.

25. In referring to the "great deposits of dynamical efficiency which are laid up for human use in our coal strata," Herschel creates an English calque for *effet dynamique*, J. N. P. Hachette's term for the product of weight and height. [J. N. P. Hachette, *Traité élémentaire des machines* (Paris: Klostermann, 1811), 1.] For a discussion of the development of this tradition, see chapter 7. The immediate influence on Herschel may well have been his reading of Charles Babbage. Herschel and Babbage had cooperated in founding the Analytical Society at Cambridge, and had collaborated on a translation of Lacroix's *Elementary Treatise on the Differential Calculus* (1816); it seems quite likely that Herschel would have glanced at Babbage's *Economy of Machines and Manufactures*.

26. Thomas Carlyle, *The Works of Thomas Carlyle*, 30 vols. (New York: Scribner's, 1896), 1:56.

27. Samuel Smiles, *The Life of George Stephenson, Railway Engineer*, 2nd ed. (London: Murray, 1857), 484–85.

28. The metaphor of "bottled" light seems to have been central to Stephenson's fascination with the idea. In a later edition, Smiles recounts how Stephenson examined an especially fat pig in a Belgian butcher-shop and inquired into its feeding. Mr. Fearon, who accompanied him, reports that "George went off into his favorite theory of the sun's light, which he said had fattened the pig; for the light had gone into the pease, and the pease had gone into the fat, and the fat pig was like a field of coal in this respect, that they were, for the most part, neither more nor less than bottled sunshine." Samuel Smiles, *The Life of George Stephenson and of his Son Robert Stephenson* (New York: Harper, 1868), 468.

29. William R. Grove, *On the Correlation of Physical Forces* (London: London Institution, 1846), 48.

30. Grove, *Physical Forces*, 5.

31. One exception to this is an 1861 article on "Force" by James Hinton, which argues that "the sum total of all the forces in the universe is equal to—nothing." "Physical action is an 0—nonentity—analysed, as it were, and spread out over time and space. Much, infinitely much, to us: nothing in itself." The article is fascinating, but it cannot be said to represent a typical Victorian response. James Hinton, "Force," 420.

32. William Benjamin Carpenter, "The Phasis of Force," *National Review* 4 (1857): 390.

33. Carpenter, *Phasis of Force*, 390.

34. Herbert Spencer, for instance, also remarks that "George Stephenson was one of the first to recognize the fact that the force impelling his locomotive, originally emanated from the sun." Spencer, *First Principles*, 283.

35. Herschel, *A Treatise of Astronomy*, 211.

36. J. F. W. Herschel, *Outlines of Astronomy* (London: Longman, Brown, 1851), 237.

37. J. F. W. Herschel, *Outlines of Astronomy* (London: Longmans, 1893), 260.

38. Smiles, *The Life of George Stephenson, Railway Engineer* (1857), 485. The sentence has been removed from Samuel Smiles, *The Life of George Stephenson, and of his Son Robert Stephenson* (1868), 468—and also from subsequent editions.

39. John Tyndall, *Heat Considered as a Mode of Motion* (New York: D. Appleton, 1864), 446.

40. Tyndall, *Heat*, 446n.

41. Tyndall's "Belfast Address," delivered before the British Association for the Advancement of Science in 1874, when Tyndall was president of that body, set forth the claims of scientific naturalism in uncompromising terms that sparked fierce controversy. See John Tyndall, *Fragments of Science* (New York: D. Appleton, 1892), 2:197. For a description of the furor that followed the address see A. S. Eve and C. H. Creasy, *The Life and Work of John Tyndall* (London: Macmillan, 1945), 186–88.

42. Tyndall, *Heat*, 447.

43. J. T. Merz, *A History of European Thought in the Nineteenth Century*, 3 vols. (Edinburgh: William Blackwood and Sons, 1912), 2:57.

44. Adams, *Degradation*, 143.

45. F. Max Müller, *Lectures on the Origin and Growth of Religion* (New York: Charles Scribner, 1879), 200. I owe this citation to Gillian Beer, " 'The Death of the Sun':Victorian Solar Physics and Solar Myth," *The Sun is God*, ed. J. B. Bullen (Oxford: Oxford University Press, 1989), 164. Beer suggests that Müller is referring to Tyndall's *Six Lectures on Light* (1878). *Heat as a Mode of Motion* seems an equally likely source for these ideas.

46. "The Source of Labour," *Chambers's Journal* series. 4, 3 (1866): 555.

47. Tyndall was a devoted reader of Carlyle, and befriended him in later life. This is only one instance of a surprisingly strong connection between Carlyle and late-Victorian naturalism; Frank Turner has pointed out that a large number of Victorian scientists, or popularizers of science, paid tribute to the influence of Carlyle. (The list includes T. H. Huxley, Francis Galton, John Morley, and Herbert Spencer as well as John Tyndall.) See Frank M. Turner, "Victorian Scientific Naturalism and Thomas Carlyle," *Victorian Studies* 18.3 (1975): 329. It would be easy to show that Carlyle's broader fascination with the transcendental significance of force was important to the popularization of the first law of thermodynamics; I have focused chapter Eight on the influence of Carlyle's productivism in order to make a more pointedly economic argument.

48. John Tyndall, "Personal Recollections of Thomas Carlyle," *New Fragments* (New York: D. Appleton, 1892), 386.

49. Rabinbach, *The Human Motor*, 84–300.

50. Bruce Clarke, *Energy Forms*. Mark Seltzer, *Bodies and Machines* (New York: Routledge, 1992).

51. Bruce Clarke, "From Thermodynamics to Virtuality," *From Energy to Information: Representation in Science and Technology, Art, and Literature*, ed. Bruce Clarke and Linda Dalrymple Henderson (Stanford: Stanford University Press, 2002), 17–33.

52. S. Dillon Ripley, "Our Onlie Begetter," *Fire of Life: The Smithsonian Book of the Sun* (New York: Smithsonian Exposition Books, 1981), 17.

BIBLIOGRAPHY

Abrams, M. H. *The Mirror and the Lamp: Romantic Theory and the Critical Tradition.* New York: Norton, 1958.

Adams, Henry. *The Degradation of the Democratic Dogma.* New York: Macmillan, 1919.

All the Year Round. "Force and Matter." (July 21, 1866): 35–38.

Amos, Andrew. "Shelley and His Contemporaries at Eton." *Athenaeum* (April 15, 1848): 390.

Arendt, Hannah. *The Human Condition.* Chicago: University of Chicago Press, 1958.

Aristotle. *Metaphysics.* Loeb Classical Library. 2 vols. Cambridge: Harvard University Press, 1936.

———. *The Art of Rhetoric.* Loeb Classical Library. Cambridge: Harvard University Press, 1939.

Babbage, Charles. *On the Economy of Machinery and Manufactures.* London: Charles Knight, 1832.

———"On the General Principles which Regulate the Application of Machinery to Manufacture and the Mechanical Arts." In *Encyclopaedia Metropolitana,* vol. 8. London: B. Fellowes, 1845.

Bachelard, Gaston. *The Psychoanalysis of Fire.* Trans. Alan C. M. Ross. Boston: Beacon Press, 1964.

Beddoes, Thomas. "Observations on the Character and Writings of John Brown, M. D." In *The Elements of Medicine,* by John Brown, vol. 1, xxxv–clxviii. London: J. Johnson, 1795.

———. *Observations on the Nature and Cure of Calculus, Sea Scurvy, Consumption, Catarrh, and Fever: Together with Conjectures upon Several Other Subjects of Physiology and Pathology.* Philadelphia: T. Dobson, 1797.

Beer, Gillian. " 'The Death of the Sun': Victorian Solar Physics and Solar Myth." In *The Sun is God,* ed. J. B. Bullen, 159–80. Oxford: Oxford University Press, 1989.

Benis, Toby R. *Romanticism on the Road: The Marginal Gains of Wordsworth's Homeless.* New York: St. Martin's, 2000.

Bensaude-Vincent, Bernadette. *Lavoisier: Mémoires d'une révolution.* Paris: Flammarion, 1993.

Berthollet, Claude Louis. "De l'influence de la lumière." *Journal de physique, de chimie, d'histoire naturelle et des arts* 29 (1786): 81–85.

Bewell, Alan. "Jefferson's Thermometer: Colonial Biogeographical Constructions of the Climate of America." In *Romantic Science: The Literary Forms of Natural History,* ed. Noah Heringman, 111–38. Albany, NY: SUNY Press, 2003.

Bilsborrow, Dewhurst. "To Erasmus Darwin, on his Work Entitled Zoonomia." In *Zoonomia; or, the Laws of Organic Life*, by Erasmus Darwin, 1:vii-viii. 2 vols. London: J. Johnson, 1794.

Blake, William. *The Complete Poetry and Prose*. Ed. David V. Erdman. Berkeley: University of California Press, 1982.

Bloom, Harold. *The Visionary Company: A Reading of English Romantic Poetry*. New York: Doubleday, 1963.

Boerhaave, Hermann. *Elements of Chemistry, Being the Annual Lectures of Herman Boerhaave, M. D.* London: J. & J. Pemberton, 1735.

Boswell, James. *Boswell's Life of Johnson*. Ed. G. B. Hill and L. F. Powell. Oxford: Oxford University Press, 1971.

Bourdieu, Pierre. "What Makes a Social Class? On the Theoretical and Practical Existence of Groups." *Berkeley Journal of Sociology* 32 (1987): 1–17.

———. *The Field of Cultural Production: Essays on Art and Literature*. New York: Columbia University Press, 1993.

Boyd, Henry. "The Helots." In *Poems, Chiefly Dramatic and Lyric*. Dublin: Graisberry and Campbell, 1793.

British Encyclopedia; or, Dictionary of the Arts and Sciences. London: Longman, Hurst, Rees, 1809–1811.

Brooke, Henry. *The Fool of Quality; or, the History of Henry Earl of Moreland*. 2nd ed. 5 vols. London: W. Johnston, 1767–1770.

Brown, John. *The Elements of Medicine*. Ed. Thomas Beddoes. 2 vols. London: J. Johnson, 1795.

Bryant, Jacob. *A New System, or an Analysis of Ancient Mythology*. 3 vols. New York: Garland, 1979. First published 1775–1776.

Burke, Edmund. "A Letter to John Farr and John Harris, Esqrs., Sheriffs of the City of Bristol, on the Affairs of America." In *The Works of the Right Honorable Edmund Burke*, vol. 2, 189–245. 12 vols. London: George Bell & Sons, 1875–1876.

———. *A Philosophical Inquiry into the Origin of our Ideas of the Sublime and Beautiful*. Ed. J. T. Boulton. New York: Columbia University Press, 1958.

Burkert, Walter. *Greek Religion*. Trans. John Raffan. Cambridge: Harvard University Press, 1985.

Butler, Marilyn. "Romantic Manichaeism: Shelley's 'On the Devil, and Devils' and Byron's Mythological Dramas." In *The Sun is God: Painting, Literature, and Mythology in the Nineteenth Century*, ed. J. B. Bullen, 13–37. Oxford: Oxford University Press, 1989.

———. *Romantics, Rebels, and Reactionaries: English Literature and its Background 1760–1830*. Oxford: Oxford University Press, 1990.

Butter, Peter. *Shelley's Idols of the Cave*. Edinburgh: Edinburgh University Press, 1954.

Byron, George Gordon, sixth Lord. *Letters and Journals*. Ed. Leslie A. Marchand. 12 vols. London: Murray, 1973–1982.

———. *The Complete Poetical Works*. Ed. Jerome McGann. 6 vols. Oxford: Oxford University Press, 1986.

Cantor, Geoffrey. "Faraday's Search for the Gravelectric Effect." *Physics Education* 26 (1991): 289–93.

Cardwell, D. S. L. *From Watt to Clausius: The Rise of Thermodynamics in the Early Industrial Age.* Ames: Iowa State University Press, 1989.

Carlyle, Thomas. *The Works of Thomas Carlyle.* 30 vols. New York: Scribner's, 1896.

Carlyon, Clement. *Early Years and Late Reflections.* London: Whittaker, 1858.

Carnot, Lazare N. M. *Principes fondamentaux de l'equilibre et du mouvement.* Paris: Bachelier, 1803.

Carpenter, J. "What We Owe to the Sun." *Once a Week* 15 (1866): 410–14.

Carpenter, William Benjamin. "The Phasis of Force." *National Review* 4 (1857): 359–93.

Cartwright, John. *Take Your Choice!* London: J. Almon, 1776.

Chalker, John. *The English Georgic: A Study in the Development of a Form.* Baltimore: Johns Hopkins Press, 1969.

Christensen, Jerome. *Lord Byron's Strength: Romantic Writing and Commercial Society.* Baltimore: Johns Hopkins University Press, 1993.

Christmas, William J. *The Lab'ring Muses: Work, Writing, and the Social Order in English Plebeian Poetry, 1730–1830.* Newark: University of Delaware Press, 2001.

Clare, John. *The Later Poems of John Clare.* Ed. Eric Robinson and David Powell. Oxford: Oxford University Press, 1984.

———. *The Early Poems of John Clare: 1804–1822.* Ed. Eric Robinson and David Powell. Oxford: Oxford University Press, 1989.

Clarke, Bruce. *Energy Forms: Allegory and Science in the Era of Classical Thermodynamics.* Ann Arbor: University of Michigan Press, 2001.

———. *From Energy to Information: Representation in Science and Technology, Art, and Literature.* Ed. Bruce Clarke and Linda Dalrymple Henderson. Stanford: Stanford University Press, 2002.

Cohen, I. Bernard. *Franklin and Newton. Memoirs of the American Philosophical Society.* Vol. 43. Philadelphia: American Philosophical Society, 1956.

Colden, Cadwallader. *The Principles of Action in Matter.* London: R. Dodsley, 1751.

Coleridge, Samuel Taylor. *Collected Letters.* Ed. E. L. Griggs. 6 vols. Oxford: Oxford University Press, 1956.

———. *The Notebooks of Samuel Taylor Coleridge.* Ed. Kathleen Coburn. 4 vols. New York: Pantheon, 1957.

———. *Complete Poetical Works.* Ed. E. H. Coleridge. 2 vols. Oxford: Oxford University Press, 1968.

———. Letter 27c/6, ms., Humphry Davy Papers. Royal Institution, London.

Connell, Phillip. *Romanticism, Economics, and the Question of "Culture."* Oxford: Oxford University Press, 2001.

Coriolis, G. G. *Du calcul de l'effet des machines, ou considérations sur l'emploi des moteurs et sur leur évaluation, pour servir d'introduction à l'étude spéciale des machines.* Paris: Carilian-Goery, 1829.

Cowper, William. *The Letters and Prose Writings of William Cowper.* Ed. J. King and C. Ryskamp. 5 vols. Oxford: Oxford University Press, 1979–1986.

———. *The Poems of William Cowper.* Ed. J. D. Baird and C. Ryskamp. 3 vols. Oxford: Oxford University Press, 1995.

Craven, Dorothy Hadley. "Cowper's Use of 'Slight Connection' in The Task: a Study in Structure and Style." PhD diss., University of Colorado, 1953.

Crosland, Maurice. "The Image of Science as a Threat: Burke versus Priestley and the 'Philosophic Revolution.' " *British Journal for the History of Science* 20 (1987): 277–307.

Darnton, Robert. *Mesmerism and the End of the Enlightenment.* Cambridge: Harvard University Press, 1968.

Davidoff, Leonore and Catherine Hall. *Family Fortunes: Men and Women of the English Middle Class 1780–1850.* Chicago: University of Chicago Press, 1987.

Davy, Humphry. *Elements of Chemical Philosophy.* London: J. Johnson, 1812.

———. *The Collected Works of Humphry Davy.* Ed. John Davy. 9 vols. London: Smith, Elder & Co., 1839.

———. *The Poetry of Humphry Davy.* Ed. Alison Pritchard. Penwith: Penwith District Council, 1978.

———. Humphry Davy Papers. The Royal Institution, London.

Davy, John. *Memoirs of the Life of Sir Humphry Davy.* 2 vols. London: Longman, 1836.

Delon, Michel. *L'idée d'énergie au tournant des Lumières (1770–1820).* Paris: Presses Universitaires de France, 1988.

Desaguliers, J. T. *A Course of Experimental Philosophy.* 2 vols. London: J. Senex, 1734.

Dickinson, H. T. *Liberty and Property: Political Ideology in Eighteenth-Century Britain.* New York: Holmes and Meier, 1977.

Diderot, Denis. "Essai sur le mérite et le vertu." *Œuvres Complètes,* vol. 1. Paris: Hermann, 1975.

Dobbs, B. J. T. *The Janus Faces of Genius: The Role of Alchemy in Newton's Thought.* Cambridge: Cambridge University Press, 1991.

Drummond, William. *The Œdipus Judaicus.* London: Valpy, 1811.

Du Châtelet, Gabrielle Emilie Le Tonnelier de Breteuil, marquise. *Dissertation sur la nature et la propagation du feu.* Paris: Prault, 1744.

Dupuis, Charles-François. *Abrégé de l'origine de tous les cultes.* Paris: Bossange, 1820.

Durling, Dwight L. *Georgic Tradition in English Poetry.* New York: Columbia University Press, 1935.

Dyer, John. *The Fleece.* In *Minor Poets of the Eighteenth Century,* ed. H. Fausset, vol. 3. New York: E. P. Dutton, 1930.

Eaves, Morris. *The Counter-Arts Conspiracy: Art and Industry in the Age of Blake.* Ithaca: Cornell University Press, 1992.

Edmonds, Charles, ed. *Poetry of the Anti-Jacobin.* London: Sampson, 1890.

Ellis, Markman. *The Politics of Sensibility: Race, Gender and Commerce in the Sentimental Novel.* Cambridge: Cambridge University Press, 1996.

Erdman, David V. *The Illuminated Blake.* New York: Doubleday, 1974.

Eve, A. S. and C. H. Creasy. *The Life and Work of John Tyndall.* London: Macmillan, 1945.

Ewart, Peter. "On the Measure of Moving Force." *Memoirs of the Literary and Philosophical Society of Manchester* 2nd series 2 (1813): 105–258.

Faraday, Michael. *Experimental Researches in Chemistry and Physics.* London: Taylor & Francis, 1859.

Feingold, Richard. *Nature and Society.* New Brunswick: Rutgers University Press, 1978.

Ferber, Michael. *The Social Vision of William Blake.* Princeton, NJ: Princeton University Press, 1985.

Fielding, Henry. *The Covent-Garden Journal and a Plan of the Universal Register-Office*. Middletown:Wesleyan University Press, 1988.

———. *The History of Tom Jones, a Foundling*. Ed. Martin C. Battestin and Fredson Bowers. 2 vols. Oxford:Weslayan University Press, 1975.

Foucault, Michel. *The Order of Things: An Archaeology of the Human Sciences*. New York: Random House, 1970.

Frye, Northrop. *Fearful Symmetry: A Study of William Blake*. Princeton: Princeton University Press, 1947.

Gale,T. *Electricity, or Ethereal Fire*.Troy: Moffitt & Lyon, 1802.

Gilchrist, Anne. "The Indestructibility of Force." *Macmillan's Magazine*, 6 (1862): 337–44.

Gill, Stephen. *William Wordsworth:A Life*. Oxford: Oxford University Press, 1989.

Gillispie, Charles. *Lazare Carnot, Savant*. Princeton: Princeton University Press, 1971.

Gilpin, William. *Three Essays: On Picturesque Beauty, on Picturesque Travel, and on Sketching Landcape*. 2nd ed. London: Blamire, 1794.

Girtanner, Christoph. "Sur l'Irritabilité, considerée comme principe de vie dans la Nature organisée." *Observations sur la physique* 36 (1790): 422–40 and 37 (1790): 139–54.

Godwin, William. *Caleb Williams*. Ed. David McCracken. London: Oxford University Press, 1970.

———. *Political and Philosophical Writings*. 7 vols. London: Pickering, 1993.

Gooding, David. "Metaphysics versus Measurement: the Conversion and Conservation of Force in Faraday's Physics." *Annals of Science* 37 (1980): 1–29.

Greenblatt, Stephen. *Shakespearean Negotiations: The Circulation of Social Energy in Renaissance England*. Berkeley: University of California Press, 1988.

Griffin, Dustin. "Redefining Georgic: Cowper's *Task*." *English Literary History* 57.4 (1990): 875–79.

Grove,William R. *On the Correlation of Physical Forces*. London: London Institution, 1846.

Guillory, John. *Cultural Capital*. Chicago: University of Chicago Press, 1993.

Hachette, J. N. P. *Traité élémentaire des machines*. Paris: Klostermann, 1811.

Hales, Stephen. *Vegetable Staticks*. London: Oldbourne, 1961. First published 1727.

Hamilton, Alexander. *The Reports of Alexander Hamilton*. Ed. Jacob E. Cooke. New York: Harper, 1964.

Hamilton, Elizabeth. *Translation of the Letters of a Hindoo Rajah*. 2 vols. London: G. G. & J. Robinson, 1796.

———. *Memoirs of Modern Philosophers*. 3 vols. Bath: G. G. and J. Robinson, 1800.

Harington, John. *The Odes and Epodon of Horace*. London:W. Crooke, 1684.

Harris, John. *The Atheist's Objections, against the Immaterial Nature of God, and Incorporeal Substances, Refuted*. London: 1698.

Hartman, Geoffrey H. "Romanticism and Anti-Self-Consciousness." In *Romanticism and Consciousness*, ed. Harold Bloom, 46–56. New York:W.W. Norton, 1970.

———. *Wordsworth's Poetry 1787–1814*. Cambridge, MA: Harvard University Press, 1987.

Hayley, William. *Poems and Plays by William Hayley*. London: T. Cadell, 1788.

Hazlitt, William. *The Complete Works*. Ed. P. P. Howe. 21 vols. London: J. M. Dent, 1930.

Hegel, G. W. F. *The Philosophy of History*. Trans. J. Sibree. New York: Dover, 1956.

———. *Hegel and the Human Spirit: A Translation of the Jena Lectures on the Philosophy of Spirit (1805–6)*. Trans. Leo Rauch. Detroit: Wayne State University Press, 1983.

Heilbron, J. L. *Electricity in the Seventeenth and Eighteenth Centuries*. Berkeley: University of California Press, 1979.

Heimann, P. M. " 'Nature is a Perpetual Worker': Newton's Æther and Eighteenth-Century Natural Philosophy." *Ambix* 20 (1973): 1–25.

Heimann, P. M. and J. E. McGuire. "Newtonian Forces and Lockean Powers: Concepts of Matter in Eighteenth-Century Thought." *Historical Studies in the Physical Sciences* 3 (1971): 233–306.

Heinzelman, Kurt. *The Economics of the Imagination*. Amherst: University of Massachusetts Press, 1980.

Helmholtz, Hermann von. "On the Conservation of Force; a Physical Memoir." In *Scientific Memoirs, Selected from the Transactions of Foreign Academies of Science, and from Foreign Journals: Natural Philosophy*, ed. John Tyndall and William Francis. London: Taylor & Francis, 1853.

———. "On the Interaction of Physical Forces." Trans. John Tyndall. In *The Correlation and Conservation of Forces*, ed. Edward L. Youmans, 211–47. New York: Appleton, 1865.

Helmont, Jean Baptiste van. *Ortus Medicinae*. Amsterdam, 1648.

Herschel, J. F. W. *A Treatise of Astronomy*. London: Longman, 1833.

———. *Outlines of Astronomy*. London: Longman, Brown, 1851.

———. *Outlines of Astronomy*. London: Longmans, 1893.

Hertz, Neil H. "Wordsworth and the Tears of Adam," *Studies in Romanticism* 7 (1967): 15–33.

Hickey, Alison. *Impure Conceits: Rhetoric and Ideology in Wordsworth's Excursion*. Stanford: Stanford University Press, 1997.

Hinton, James. "Force." *Cornhill Magazine* 4 (1861): 409–20.

Holcroft, Thomas. *The Adventures of Hugh Trevor*. Ed. Seamus Deane. London: Oxford University Press, 1973.

Hölderlin, Friedrich. "Patmos." In *Poems and Fragments*, trans. Michael Hamburger, 482–507. London: Anvil, 1994.

Homer. *Odyssey*. Trans. Richmond Lattimore. New York: Harper, 1967.

Hutton, James. *Dissertations on Different Subjects in Natural Philosophy*. Edinburgh: A. Strahan and T. Cadell, 1792.

———. *A Dissertation on the Philosophy of Light, Heat, and Fire*. Edinburgh: Cadell, 1794.

Ingenhousz, Jan. *Experiments upon Vegetables*. London: P. Elmsley, 1779.

Ireland, William Henry. *Rhapsodies*. London: Longman and Rees, 1803.

———. *Neglected Genius*. London: Printed by W. Wilson for George Cowrie and Co., 1812.

Jack, Ian. *Keats and the Mirror of Art*. Oxford: Oxford University Press, 1967.

Jaton, Anne-Marie. "Énergétique et féminité (1770–1820)." *Romantisme* 46 (1984): 15–25.

Jeffrey, Francis. Review of *The Dramatic Works of John Ford*. In *Contributions to the Edinburgh Review*, ed. Henry Weber, vol. 2. London: Longmans, 1944.

Jenkins, Alice. "Humphry Davy: Poetry, Science, and the Love of Light." *1798: The Year of the Lyrical Ballads*. Ed. Richard Cronin. Houndmills: Macmillan, 1998.

Johnson, Randal. "Editor's Introduction." In *The Field of Cultural Production: Essays on Art and Literature*, by Pierre Bourdieu. New York: Columbia University Press, 1993.

Johnson, Samuel. *A Dictionary of the English Language*. London: Strahan, 1755.

Johnston, Kenneth. *The Hidden Wordsworth: Poet, Lover, Rebel, Spy*. New York: W. W. Norton, 1998.

Jonson, Ben. *Poems*. Ed. I. Donaldson. London: Oxford University Press, 1975.

Jordan, Sarah. *The Anxieties of Idleness: Idleness in Eighteenth-Century British Literature and Culture*. Lewisburg: Bucknell University Press, 2003.

Keats, John. *The Letters of John Keats*. Ed. Hyder Edward Rollins. 2 vols. Cambridge, MA.: Harvard University Press, 1958.

———. *The Poems of John Keats*. Ed. Jack Stillinger. Cambridge, MA: Harvard University Press, 1978.

Keen, Paul. *The Crisis of Literature in the 1790s: Print Culture and the Public Sphere*. Cambridge: Cambridge University Press, 1999.

Kepler, Johannes. *New Astronomy*. Trans. William H. Donahue. Cambridge: Cambridge University Press, 1992.

King, James. *William Cowper: A Biography*. Durham: Duke University Press, 1986.

Knight, Richard Payne. *The Progress of Civil Society*. London: G. Nicol, 1796.

Kramnick, Isaac. *Republicanism and Bourgeois Radicalism: Political Ideology in Late Eighteenth-Century England and America*. Ithaca: Cornell University Press, 1990.

Landon, Letitia Elizabeth. (L. E. L.) *Selected Writings*. Ed. Jerome McGann and Daniel Riess. Peterborough, Ontario: Broadview, 1997.

Larson, James L. *Interpreting Nature: The Science of Living Form from Linnaeus to Kant*. Baltimore: Johns Hopkins University Press, 1994.

Larson, Magali Sarfatti. *The Rise of Professionalism: A Sociological Analysis*. Berkeley: University of California Press, 1977.

Lavoisier, Antoine-Laurent. *Elements of Chemistry*. Trans. Robert Kerr. New York: Dover, 1965.

Lavoisier, Antoine-Laurent and Armand Seguin. "Premier mémoire sur la respiration des animaux." In *Œuvres*, by Antoine-Laurent Lavoisier, vol. 2, 688–703. Paris: Imprimerie impériale, 1862.

Leighton, Angela. *Shelley and the Sublime: An Interpretation of the Major Poems*. Cambridge: Cambridge University Press, 1984.

Liebig, Justus. "The Connection and Equivalence of Forces." In *The Correlation and Conservation of Forces*, ed. Edward L. Youmans, 387–97. New York: Appleton, 1865.

Liu, Alan. *Wordsworth: The Sense of History*. Stanford: Stanford University Press, 1989.

Locke, John. *An Essay Concerning Human Understanding*. Ed. A. C. Fraser. 2 vols. New York: Dover, 1959.

Lovett, Richard. *Philosophical Essays*. London: J. Johnson, 1766.

Lukács, Georg. "Reification and the Consciousness of the Proletariat." In *History and Class Consciousness*, trans. Rodney Livingstone, 83–222. Cambridge, MA: MIT Press, 1971.

Lundgreen, Peter. "Engineering Education in Europe and the U.S.A, 1750–1930: The Rise to Dominance of School Culture and the Engineering Professions." *Annals of Science* 47 (1990): 36–47.

Lussier, Mark. *Romantic Dynamics: The Poetics of Physicality*. New York: St. Martin's Press, 2000.

MacLurg, James. *Experiments upon the Human Bile*. London: T. Cadell, 1772.

Malthus, T. R. *An Essay on the Principle of Population*. 3rd ed. 2 vols. London: J. Johnson, 1806.

———. "Spence on Commerce." *Edinburgh Review* 11 (1808): 429–48.

———. *An Inquiry into the Nature and Progress of Rent, and the Principles by which it is Regulated*. London: Printed for John Murray and J. Johnson, 1815.

———. *The Grounds of an Opinion on the Policy of Restricting the Importation of Foreign Corn*. London: Printed for John Murray and J. Johnson, 1815.

Marx, Karl. *Economic and Philosophic Manuscripts of 1844*. Moscow: Foreign Languages Publishing House, 1961.

———. *Capital: A Critique of Political Economy*. Vol. 1. Trans. Ben Fowkes. New York: Vintage, 1977.

Marx, Leo. *The Machine in the Garden: Technology and the Pastoral Ideal in America* Oxford: Oxford University Press, 2000. First published 1964.

McGann, Jerome. *Fiery Dust: Byron's Poetic Development*. Chicago: University of Chicago Press, 1968.

———. *The Poetics of Sensibility*. Oxford: Oxford University Press, 1996.

———. "Byron, Mobility, and the Poetics of Historical Ventriloquism." In *Byron and Romanticism*, by Jerome McGann, ed. James Soderholm, 36–52. Cambridge: Cambridge University Press, 2002.

McGuire, J. E. "Force, Active Principles, and Newton's Invisible Realm." *Ambix* 15 (1968): 154–208.

Merz, J. T. *A History of European Thought in the Nineteenth Century*. 3 vols. Edinburgh: William Blackwood and Sons, 1912.

Messman, Frank J. *Richard Payne Knight: The Twilight of Virtuosity*. Paris: Mouton, 1973.

Milton, John. *Paradise Lost*. Ed. Scott Elledge. New York: Norton, 1975.

Mirowski, Philip. *More Heat than Light: Economics as Social Physics, Physics as Nature's Economics*. Cambridge: Cambridge University Press, 1989.

Mitchell, W. J. T. "Poetic and Pictorial Imagination in Blake's *The Book of Urizen*." *Eighteenth Century Studies* 3 (1969): 83–107.

Moore, C. A. "Whig Panegyric Verse, 1700–1760." *PMLA* 41 (1926): 362–401.

Moore, Thomas. *Poetical Works*. 10 vols. London: Orme, Brown, Green, and Longmans, 1840–1841.

Morton, Alan Q. "Concepts of Power: Natural Philosophy and the Uses of Machines in Mid-Eighteenth-Century London." *British Journal for the History of Science* 28 (1995): 63–78.

Muir, John. *My First Summer in the Sierra*. Boston: Houghton Mifflin, 1911.

Müller, F. Max. *Lectures on the Origin and Growth of Religion*. New York: Charles Scribner, 1879.

Nash, Leonard K. *Plants and the Atmosphere*. Cambridge: Harvard University Press, 1952.

Newton, Isaac. *Opticks; or, A Treatise of the Reflections, Refractions, Inflections and Colours of Light*. New York: Dover, 1952.

―――. "Of Natures Obvious Laws & Processes in Vegetation." In *The Janus Faces of Genius: The Role of Alchemy in Newton's Thought*, by B. J. T. Dobbs, 256–70. Cambridge: Cambridge University Press, 1991.

Ovid. *Metamorphoses*. Trans. Rolfe Humphries. Bloomington: Indiana University Press, 1955.

Patterson, Annabel. "Wordsworth's Georgic: Genre and Structure in *The Excursion*." *The Wordsworth Circle* 9 (1978): 145–54.

Peacock, Thomas Love. *Works*. 10 vols. London: Constable, 1924–1934.

Peterfreund, Stuart. "The Re-Emergence of Energy in the Discourse of Literature and Science." *Annals of Scholarship: Metastudies of the Humanities and Social Sciences* 4 (1986–1987): 22–53.

―――. *William Blake in a Newtonian World: Essays on Literature as Art and Science*. Series for Science and Culture. Norman: University of Oklahoma Press, 1998.

Pfau, Thomas. *Wordsworth's Profession: Form, Class, and the Logic of Early Romantic Cultural Production*. Stanford, California: Stanford University Press, 1997.

Pocock, J. G. A. *The Machiavellian Moment: Florentine Political Thought and the Atlantic Republican Tradition*. Princeton, NJ: Princeton University Press, 1975.

―――. *Virtue, Commerce, and History: Essays on Political Thought and History, Chiefly in the Eighteenth Century*. Cambridge: Cambridge University Press, 1985.

Poirier, Jean-Pierre. *Lavoisier: Chemist, Biologist, Economist*. Trans. Rebecca Balinski Philadelphia: University of Pennsylvania Press, 1996.

Porter, Theodore M. *Trust in Numbers: The Pursuit of Objectivity in Science and Public Life*. Princeton: Princeton University Press, 1995.

Prest, Wilfrid. "The Professions and Society in Early Modern England." In *The Professions in Early Modern England*, ed. Wilfrid Prest, 1–24. London: Croom Helm, 1987.

Price, Martin. *To the Palace of Wisdom: Studies in Order and Energy from Dryden to Blake*. New York: Doubleday, 1964.

Priestley, Joseph. *Experiments and Observations on Different Kinds of Air*. 3 vols. London: J. Johnson, 1774–1777.

―――. *Disquisitions Relating to Matter and Spirit*. London: J. Johnson, 1777.

Priestman, Martin. *Cowper's Task: Structure and Influence*. Cambridge: Cambridge University Press, 1983.

Project for American and French Research on the Treasury of the French Language. University of Chicago, World-Wide Web < http://humanities.uchicago.edu/ARTFL/ARTFL.html/ > (accessed October 1995).

Punter, David. "Blake: Creative and Uncreative Labor." *Studies in Romanticism* 16 (1977): 535–61.

Rabinbach, Anson. *The Human Motor: Energy, Fatigue, and the Origins of Modernity.* New York: Basic Books, 1990.

Radcliffe, Ann. *The Mysteries of Udolpho.* Ed. Bonamy Dobrée. New York: Oxford University Press, 1992.

Rankine, William John Macquorn. "On the General Law of the Transformation of Energy." *Philosophical Magazine* 4th series 5 (1853): 106–17.

Reader, W. J. *Professional Men: The Rise of the Professional Classes in Nineteenth-Century England.* London: Weidenfeld and Nicolson, 1966.

Reynolds, Terry S. *Stronger than a Hundred Men: A History of the Vertical Water Wheel.* Baltimore: Johns Hopkins University Press, 1983.

Ricardo, David. *On the Principles of Political Economy and Taxation.* Vol. 1. In *Works and Correspondence,* ed. Pierro Sraffa and M. H. Dobb. 10 vols. Cambridge: Cambridge University Press, 1966.

Ripley, S. Dillon. "Our Onlie Begetter." In *Fire of Life: The Smithsonian Book of the Sun,* 13–17. New York: Smithsonian Exposition Books, 1981.

Risse, Guenter B. "Brunonian Therapeutics: New Wine in Old Bottles?" In *Brunonianism in Britain and Europe,* ed. W. F. Bynum and Roy Porter. London: Wellcome Institute, 1988.

Robertson, John. "The Scottish Enlightenment at the Limits of the Civic Tradition." In *Wealth and Virtue: The Shaping of Political Economy in the Scottish Enlightenment,* ed. Istvan Hont and Michael Ignatieff. Cambridge: Cambridge University Press, 1983.

Robinson, Mary. *The Poetical Works of the Late Mrs. Mary Robinson.* 3 vols. London: Richard Phillips, 1806.

Roe, Nicholas. " 'Atmospheric Air Itself': Medical Science, Politics, and Poetry in Thelwall, Coleridge, and Wordsworth." In *1798: The Year of the Lyrical Ballads,* ed. Richard Cronin, 185–202. Houndmills: Macmillan, 1998.

Ross, Trevor. " 'Pure Poetry': Cultural Capital and the Rejection of Classicism." *Modern Language Quarterly* 58 (1997): 437–56.

Rousseau, Jean-Jacques. *La nouvelle Héloïse.* Paris: Hachette, 1925.

Russell, James B. "The Labour of the Sunbeams." *Recreative Science* 3 (1862): 56–60.

Rzepka, Charles. " 'A Gift that Complicates Employ': Poetry and Poverty in 'Resolution and Independence.' " *Studies in Romanticism* 28 (1989): 225–47.

Say, Jean-Baptiste. *Traité d'Economie Politique.* Paris: Deterville, 1803.

Schaffer, Simon. "States of Mind: Enlightenment and Natural Philosophy." In *The Languages of Psyche,* ed. G. S. Rousseau, 233–90. Berkeley and Los Angeles: University of California Press, 1990.

Schiller, Friedrich. *On the Aesthetic Education of Man: In a Series of Letters.* Ed. and trans. Elizabeth M. Wilkinson and L. A. Willoughby. Oxford: Oxford University Press, 1967.

Schlegel, Friedrich. *Dialogue on Poetry and Literary Aphorisms.* Trans. Ernst Behler and Roman Struc. University Park: Pennsylvania State University Press, 1968.

Schofield, Robert. *Mechanism and Materialism: British Natural Philosophy in an Age of Reason.* Princeton: Princeton University Press, 1970.

Schumpeter, Joseph A. *History of Economic Analysis.* Ed. Elizabeth Boody Schumpeter. New York: Oxford University Press, 1954.

Seltzer, Mark. *Bodies and Machines*. New York: Routledge, 1992.

Senebier, Jean. *Mémoires physico-chymiques sur l'influence de la lumière solaire pour modifier les êtres des trois règnes de la nature, et sur-tout ceux du règne végétal*. 3 vols. Geneva: Chiral, 1782.

———. *Physiologie végétale, contenant une description des organes des plantes*. Geneva: 1799–1800.

———. *Essai sur l'art d'observer et de faire des expériences*. 2nd ed. 3 vols. Geneva: 1802.

Senior, Nassau W. *An Outline of the Science of Political Economy*. London: Clowes, 1836.

Seward, Anna. *The Poetical Works of Anna Seward*. Ed. Walter Scott. 3 vols. Edinburgh: J. Ballantyne, 1810.

Shapin, Steven. "Social Uses of Science." In *The Ferment of Knowledge: Studies in the Historiography of Eighteenth-Century Science*, ed. G. S. Rousseau and Roy Porter, 93–139. Cambridge: Cambridge University Press, 1980.

Shelley, Mary. *Proserpine and Midas*. Ed. A Koszul. London: Humphrey Milford, 1922.

———. *Journal*. Ed. F. L. Jones. Norman: University of Oklahoma Press, 1947.

Shelley, Percy Bysshe. *The Letters of Percy Bysshe Shelley*. Ed. Frederick L. Jones. 2 vols. Oxford: Oxford University Press, 1964.

———. *Poetical Works*. Ed. Thomas Hutchinson, new edition corrected by G. M. Matthews. London: Oxford University Press, 1970.

———. *Shelley's Prose; or, The Trumpet of a Prophecy*. Ed. David Lee Clark. London: Fourth Estate, 1988.

Sidney, Sir Philip. *Complete Works*. Ed. A. Feuillerat. 4 vols. Cambridge: Cambridge University Press, 1923.

Simpson, David. *Wordsworth's Historical Imagination: The Poetry of Displacement*. New York: Methuen, 1987.

Siskin, Clifford. *The Work of Writing: Literature and Social Change in Britain, 1700–1830*. Baltimore: The Johns Hopkins University Press, 1998.

Smeaton, John. *An Experimental Enquiry Concerning the Natural Powers of Water and Wind to Turn Mills, and Other Machines, Depending on a Circular Motion*. London, 1760.

———. *An Experimental Inquiry . . . and an Experimental Examination of the Quantity and Proportion of Mechanic Power Necessary to be Employed in Giving Different Degrees of Velocity to Heavy Bodies from a State of Rest*. London: Taylor, 1796. First published 1776.

Smiles, Samuel. *The Life of George Stephenson, Railway Engineer*. 2nd ed. London: Murray, 1857.

———. *The Life of George Stephenson and of his Son Robert Stephenson*. New York: Harper, 1868.

Smith, Adam. *An Inquiry into the Nature and Causes of the Wealth of Nations*. Ed. R. H. Campbell, A. S. Skinner, and W. B. Todd. 2 vols. Oxford: Oxford University Press, 1976.

Smith, Charlotte. *Elegiac Sonnets*. 2 vols. London: T. Cadell and W. Davies, 1797–1800.

Smith, Crosbie. *The Science of Energy: A Cultural History of Energy Physics in Victorian Britain*. Chicago: University of Chicago Press, 1998.

Solvay, Ernest. *Principes d'Orientation Sociale*. Bruxelles: Misch et Thron, 1904.

"The Source of Labour." *Chambers's Journal* series 4, 3 (1866): 554–57.

Southey, Robert. "The Peruvian's Dirge over the Body of his Father." In *Poetical Works*, 10 vols, 2:207–09. London: Longmans, 1838.

Spence, William. *Agriculture the Source of the Wealth of Britain*. First published 1808, reprinted in William Spence, *Tracts on Political Economy*. London: Longman, Hurst, Rees, Orme, and Brown, 1822.

Spencer, Herbert. *First Principles*. London: Williams and Norgate, 1862.

"Sun Force and Earth Force." *The Popular Science Review* 5 (1866): 327–36.

Thelwall, John. *An Essay Towards a Definition of Animal Vitality*. London, 1793.

Thomson, James. *The Complete Poetical Works of James Thomson*. Ed. J. Logie Robertson. London: Oxford University Press, 1908.

Thomson, William. *Mathematical and Physical Papers*. 6 vols. Cambridge: Cambridge University Press, 1882–1911.

Thomson, William and P. G. Tait. "Energy." Good Words 3 (1862): 601–07.

Thoreau, Henry David. *Walden and Resistance to Civil Government*. Ed. William Rossi and Owen Thomas. New York: W. W. Norton, 1992.

Trent, Joseph. *An Inquiry into the Effects of Light in Respiration*. Philadelphia: Way and Graff, 1800.

Turner, Frank M. "Victorian Scientific Naturalism and Thomas Carlyle." *Victorian Studies* 18 (1975): 325–43.

Tyndall, John. *Heat Considered as a Mode of Motion*. New York: D. Appleton, 1864.

———. *Fragments of Science*. New York: D. Appleton, 1892.

———. "Personal Recollections of Thomas Carlyle." *New Fragments*. New York: D. Appleton, 1892.

Underwood, Ted. "Productivism and the Vogue for 'Energy' in Late Eighteenth-Century Britain." *Studies in Romanticism* 34 (1995): 103–25.

———. "Romantic Historicism and the Afterlife." *PMLA* 117 (2002): 237–51.

Virgil. *The Georgics*. Trans. L. P. Wilkinson. New York: Penguin, 1982.

Wahrman, Dror. *Imagining the Middle Class: The Political Representation of Class in Britain, c. 1780–1840*. Cambridge: Cambridge University Press, 1995.

Walker, Adam. *A System of Familiar Philosophy in Twelve Lectures*. 2 vols. London: Printed for the author, 1802.

Wallace, Anne D. *Walking, Literature, and English Culture: The Origins and Uses of Peripatetic in the Nineteenth Century*. Oxford: Oxford University Press, 1993.

Wallech, Stephen. " 'Class Versus Rank': The Transformation of Eighteenth-Century English Social Terms and Theories of Production." *Journal of the History of Ideas* 47 (1986): 409–31.

Waller, Bryan Procter. (Barry Cornwall). *The Flood of Thessaly, The Girl of Provence, and Other Poems*. London: Henry Colburn, 1823.

Wasserman, Earl. *Shelley: A Critical Reading*. Baltimore: Johns Hopkins University Press, 1971.

Waterhouse, Benjamin. *On the Principle of Vitality*. Boston: Fleet, 1790.

Wiffen, Jeremiah Holmes. *Aonian Hours, and Other Poems*. London: Longman, Hurst, Rees, Orme, and Brown, 1819.

Williams, Raymond. *The Country and the City*. New York: Oxford University Press, 1975.

———. *Culture and Society: 1780–1950*. New York: Columbia University Press, 1983.

———. *Keywords: A Vocabulary of Culture and Society*. New York: Oxford University Press, 1983.

Wilson, Alexander. "An Answer to the Objections Stated by M. De La Lande." *Philosophical Transactions* 73 (1783): 144–68.

Wilson, Eric G. *The Spiritual History of Ice: Romanticism, Science, and the Imagination*. New York: Palgrave, 2003.

Wimsatt, W. K. "The Structure of Romantic Nature Imagery." In *The Verbal Icon*, by W. K. Wimsatt with the collaboration of Monroe Beardsley, 103–16. N.p.: University of Kentucky, 1954.

Wise, M. Norton, with the collaboration of Crosbie Smith. "Work and Waste: Political Economy and Natural Philosophy in Nineteenth-Century Britain." *History of Science* 27 (1989): 263–301, 391–449, and 28 (1990): 221–61.

Wollstonecraft, Mary. *A Vindication of the Rights of Woman*. Ed. Carol H. Poston. New York: Norton, 1975.

Wordsworth, William. *Poetical Works*. Ed. E. de Selincourt and Helen Darbishire. 2nd ed. 5 vols. Oxford: Oxford University Press, 1965–1966.

———. *Prose Works*. Ed. W. J. B Owen and Jane Worthington Smyser, 3 vols. Oxford: Oxford University Press, 1974.

———. *The Prelude: 1799, 1805, 1850*. Ed. Jonathan Wordsworth, M. H. Abrams, and Stephen Gill. New York: W. W. Norton, 1979.

———. *The Ruined Cottage and The Pedlar*. Ed. James Butler. Ithaca: Cornell University Press, 1979.

Wordsworth, William and Dorothy Wordsworth. *The Letters of William and Dorothy Wordsworth: The Middle Years*. Ed. Ernest de Selincourt, Mary Moorman, and Alan G. Hill. 2 vols. Oxford: Oxford University Press, 1970.

———. *The Letters of William and Dorothy Wordsworth: The Later Years*. Ed. Alan G. Hill and Ernest de Selincourt. 2nd ed. 2 vols. Oxford: Oxford University Press, 1978.

———. *The Letters of William and Dorothy Wordsworth: A Supplement of New Letters*. Ed. Alan G. Hill. New York: Oxford University Press, 1993.

Yolton, John W. *Thinking Matter: Materialism in Eighteenth-Century Britain*. Minneapolis: University of Minessota Press, 1983.

Youmans, Edward L. *The Correlation and Conservation of Forces*. New York: Appleton, 1865.

Young, Robert. *An Essay on the Powers and Mechanism of Nature*. London: T. Becket, 1788.

INDEX

.

Printed in the United States
By Bookmasters